International Conference on Information Technology and Digital Applications 2016

IOP Conference Series: Materials Science and Engineering
Volume 185

Yogyakarta, Indonesia
14 – 16 November 2016

Editors:

Taufiq Hidayat
Ari Sujarwo

ISBN: 978-1-5108-3841-3
ISSN: 1757-8981

Printed from e-media with permission by:

Curran Associates, Inc.
57 Morehouse Lane
Red Hook, NY 12571

Some format issues inherent in the e-media version may also appear in this print version.

Printed by Curran Associates, Inc. (2017)

For permission requests, please contact the Institute of Physics
at the address below.

Institute of Physics
Dirac House, Temple Back
Bristol BS1 6BE UK

Phone: 44 1 17 929 7481
Fax: 44 1 17 920 0979

techtracking@iop.org

Additional copies of this publication are available from:

Curran Associates, Inc.
57 Morehouse Lane
Red Hook, NY 12571 USA
Phone: 845-758-0400
Fax: 845-758-2633
Email: curran@proceedings.com
Web: www.proceedings.com

Table of contents

Volume 185

The International Conference on Information Technology and Digital Applications
14–16 November 2016, Yogyakarta, Indonesia

Accepted papers received: 28 February 2017
Published online: 22 March 2017

Preface

Keynote

Applications

Software Engineering

Hard Science and Engineering

Yogyakarta, 15 November 2016

Welcome to ICITDA, the Annual International Conference on Information Technology and Digital Applications 2016. On behalf of the committees, we are grateful to have your paper published in the 1st ICITDA. This International Conference is held by Department of Informatics, Faculty of Industrial Technology, Universitas Islam Indonesia, Yogyakarta, Indonesia in collaboration with Rajamangala University of Technology Thanyaburi, Thailand, and Youngsan University, South Korea.

ICITDA 2016 received 58 papers from 8 countries, and 30 papers were accepted in various topics.

Our conference will feature Prof. Wenny Rahayu, the Head of School of Engineering and Mathematical Sciences, La Trobe University, Australia and Prof. Dr. Lee Jin Ro, from the Department of Advertising and Public Relations, Youngsan University, South Korea as keynote speakers.

This year, ICITDA held in a city of tourism, Yogyakarta, Indonesia. We hope you enjoy your academics visit to this wonderful place. See you again in next ICITDA.

Best regards,

Taufiq Hidayat, ST., MCS

Conference Chair

CHAIRS

Host

- Department of Informatics, Faculty of Industrial Technology, Universitas Islam Indonesia

Editors

- Taufiq Hidayat
- Ari Sujarwo

Organizing Committee

- Taufiq Hidayat
- Septia Rani
- Ari Sujarwo
- Ahmad M. Raf'ie Pratama

Program Committee

- Prof. Wenny Rahayu – La Trobe University, Australia
- Prof. Th.P. van der Weide – Radboud University Nijmegen, The Netherlands
- Prof. Reyer Zwiggelaar – Aberystwyth University, UK
- Prof. Lori L. Scarlatos – Stony Brook University, USA
- Prof. Dr. Mohd Fadzil bin Hassan – Universiti Teknologi Petronas, Malaysia
- Prof. Dr.Ing.Ir. Iping Supriana Suwardi – Bandung Institute of Technology, Indonesia
- Prof. Dr.Ing.Ir. Benhard Sitohang – Bandung Institute of Technology, Indonesia
- Prof. Dr.Ir. Mauridhi Hery Purnomo M.Eng. – Sepuluh Nopember Institute of Technology, Indonesia
- Prof. Dr. Sri Hartati – Gadjah Mada University, Indonesia
- Assoc. Prof. Dr. Tutut Herawan, University of Malaya, Malaysia
- Dr. Uraiwan Inyaem – Rajamangala University of Technology Thanyaburi, Thailand
- Dr. Burasakorn Yoosooka – Rajamangala University of Technology Thanyaburi, Thailand
- Dr. Julaluk Watthananon – Rajamangala University of Technology Thanyaburi, Thailand
- Dr. Sri Kusumadewi – Universitas Islam Indonesia, Indonesia
- Dr. Fathul Wahid – Universitas Islam Indonesia, Indonesia

IOP Conf. Series: Materials Science and Engineering **185** (2017) 011001 doi:10.1088/1757-899X/185/1/011001

- Dr. Teduh Dirgahayu – Universitas Islam Indonesia, Indonesia
- Dr. Izzati Muhimmah – Universitas Islam Indonesia, Indonesia
- Dr. M. Andri Setiawan – Universitas Islam Indonesia, Indonesia
- Yudi Prayudi, S.Si, M.Kom – Universitas Islam Indonesia, Indonesia
- Affan Mahtarami, S.Kom, MT – Universitas Islam Indonesia, Indonesia
- Irving Vitra Paputungan, ST, M.Sc. – Universitas Islam Indonesia, Indonesia
- Syarif Hidayat, S.Kom, MIT – Universitas Islam Indonesia, Indonesia
- Dhomas Hatta Fudholi, ST, M.Eng – Universitas Islam Indonesia, Indonesia
- Ing. Ridho Rahmadi, ST, M.Sc – Universitas Islam Indonesia, Indonesia
- Kurniawan Dwi Irianto, ST, M.Sc – Universitas Islam Indonesia, Indonesia

Keynote Speeches

Prof. Wenny Rahayu

Head of School of Engineering and Mathematical Sciences, La Trobe University, Australia

Privacy Preservation in Big Databases

Advances in data collection techniques, from web application oriented through to automated sensor devices and crowd sourcing mechanisms, have significantly increased the volume and complexity of data storage and exchanges. As we move toward fully automated, seamless, and pervasive data gathering mechanisms, there is also an increase risk in personal privacy breach or disclosure of sensitive information. In this talk, well known and state-of the art studies in privacy protection for big databases will be discussed, along with new proposals in dealing with non-traditional and externally linked databases.

Prof. Jin Ro Lee

Department of Advertising and Public Relations, Youngsan University, South Korea

Cyber Space and Digital Democracy in South Korea

Korea is a very interesting case. Korea shows rapid growth of the Internet users, and larger trade surplus in telecommunication industry with the help of government's successful information technology policy. And Koreans also made their country more democratic with active participation.

This paper analyzed the growth of the Internet and SNS in South Korea. The Internet and SNS created cyber space. They have several advantages as an effective means of communication. Cyber space is influenced by three subjects such as the government [state], the market [capital], and citizens [people].

There are two research questions. First question is how the Korean CMC industry can grow fast after its birth. Three main subjects were dealt with in this research. They are the State, the Market, and the Citizen. I divided the history of Korean CMC industry into three periods. The first formation stage is from the birth of CMC in Korea between from 1980 and 1990. The government initiated the monopolistic CMC market. Several conglomerates participated to co-operate the government. But the users are very small. The second growth stage is between from 1990 to 1995. The government also deregulated the Market with changing policy from 'appointment' to 'registration'. The companies increased investment for the possibility of wide diffusion of CMC use. The third prosperity stage is between 1995 and 2010. The government promoted the CMC market's competition with 'notice' policy. And citizens actively enjoy and apply CMC services. However, the fourth shift stage to smartphone faced several problems such as less democracy and one way communication which will weaken the creativity of the content.

Second question is what the roles of three subjects are. I examined the cyber space by the uses of digital media with three subjects. Even though the state and the market have limits to promote democracy, the citizens are expected to make the digital society more democratic. If the state tries to monitor citizen by the use of digital media, that society is called 'surveillance society'. As the government is open and transparent, the citizens' democracy will increase. The market should limit to gather and accumulate peoples' information and profile for protecting their privacy.

Cyber space is a public sphere, which is two-way, economic, and open to every people. It has various positive sides. It promotes the communication by people of political information and opinion freely and actively. It makes people form groups against the wrong-doing of big companies and keeps people's privileges. People buy goods more cheaply via e-commerce, which also helps companies lessen the expenses.

However, the cyber space has several negative sides, too. The government and companies can accumulate people's information and use it for diverse purposes. If the government uses that information to monitor and control people, such a society may become a 'surveillance society' threatening democracy as in George Orwell's '1984'. As companies also take advantage of big data to sell more products for their profit increase, so people often feel they are captivated by a lot of commercial messages, including much advertising spam-mail. The more civil society depends on the cyber space, the more it can become fragile and risky.

The digital media guarantees neither democracy nor a surveillance society. It's true that new technology gives us the opportunity to expand democracy. However, if we don't use new media rightly and positively, we may find it negative or even harmless.

Therefore, the positive use of the digital media and keeping our cyber space democratic is very important. With democracy, the Internet and SNS industry can be flourished by the cooperation of netizen, business managers, IT researchers, and politicians.

Peer review statement

All papers published in this volume of *IOP Conference Series: Materials Science and Engineering* have been peer reviewed through processes administered by the proceedings Editors. Reviews were conducted by expert referees to the professional and scientific standards expected of a proceedings journal published by IOP Publishing.

IOP Conf. Series: Materials Science and Engineering **185** (2017) 012001 doi:10.1088/1757-899X/185/1/012001

Cyber Space and Digital Democracy in South Korea

Jin Ro Lee

Professor, Department of Advertising and Public Relations,Youngsan University,Korea, +82-10-6528-7725
Email: leejr@ysu.ac.kr

Abstract. Korea is a very interesting case. Korea shows rapid growth of the Internet users, and largertrade surplus in telecommunication industry with the help of government's successfulinformation technology policy. And Koreans also made their country more democratic withactive participation.This paper analyzed the growth of the Internet and SNS in South Korea. The Internet andSNS created cyber space. They have several advantages as an effective means ofcommunication. Cyber space is influenced by three subjects such as the government [state],the market [capital], and citizens [people].There are two research questions. First question is how the Korean CMC industry can growfast after its birth. Three main subjects were dealt with in this research. They are the State,the Market, and the Citizen. I divided the history of Korean CMC industry into threeperiods. The first formation stage is from the birth of CMC in Korea between from 1980 and1990. The government initiated the monopolistic CMC market. Several conglomeratesparticipated to co-operate the government. But the users are very small. The second growthstage is between from 1990 to 1995. The government also deregulated the Market withchanging policy from 'appointment' to 'registration'. The companies increased investmentfor the possibility of wide diffusion of CMC use. The third prosperity stage is between 1995and 2010. The government promoted the CMC market's competition with 'notice' policy.And citizens actively enjoy and apply CMC services. However, the fourth shift stage tosmartphone faced several problems such as less democracy and one way communicationwhich will weaken the creativity of the content.Second question is what the roles of three subjects are. I examined the cyber space by theuses of digital media with three subjects. Even though the state and the market have limitsto promote democracy, the citizens are expected to make the digital society moredemocratic. If the state tries to monitor citizen by the use of digital media, that society iscalled 'surveillance society'. As the government is open and transparent, the citizens'democracy will increase. The market should limit to gather and accumulate peoples'information and profile for protecting their privacy.Cyber space is a public sphere, which is two-way, economic, and open to every people. Ithas various positive sides. It promotes the communication by people of politicalinformation and opinion freely and actively. It makes people form groups against thewrong-doing of big companies and keeps people's privileges. People buy goods morecheaply via e-commerce, which also helps companies lessen the expenses.However, the cyber space has several negative sides, too. The government and companiescan accumulate people's information and use it for diverse purposes. If the governmentuses that information to monitor and control people, such a society may become a'surveillance society' threatening democracy as in George Orwell's '1984'. As companies alsotake advantage of big data to sell more products for their profit increase, so people oftenfeel they are captivated by a lot of commercial messages, including much advertising spammail.The more civil society depends on the cyber space, the more it can become fragile andrisky.The digital media guarantees neither democracy nor a surveillance society. It's true thatnew technology gives us the opportunity to expand democracy. However, if we don't usenew media rightly and positively, we may find it negative or even harmless.

IOP Conf. Series: Materials Science and Engineering **185** (2017) 012001 doi:10.1088/1757-899X/185/1/012001

Therefore, thepositive use of the digital media and keeping our cyber space democratic is very important.With democracy, the Internet and SNS industry can be flourished by the cooperation ofnetizen, business managers, IT researchers, and politicians.

1. Introduction

We live both in the cyber space and in the digital society. The cyber space is evolving by digital media such as Internet and social media. Digital media produce and disseminate huge information all over the society at high speed. People can interact and form a community on the website. Many researchers observe these sides and predict that digital media will make the present digital society more democratic.

Internet as a digital media has a long history in the U.S.A, which started ARPANET, predecessor of Internet, in 1969, and public access began in 1992. Compared to U.S.A, Korea has a relatively short history of Internet. Korea began to connect the global network in 1982 and opened it officially in 1994. Concretely, the computer and communications industry, which includes telephones, mobile communication and the Internet, has made rapid progress. In South Korea, the ratio of people over 6 years old who use the Internet e-mail is 61.9 percent in 2015 (http://stat.kisdi.re.kr/). And the SNS (Social Network Service, Social Media is regarded as same) usage rate is 43.1 percent (http://stat.kisdi.re.kr/). These users have been re-born as 'netizen (network + citizen)' by the advancement of information technology (IT). They play a greater role in Korean digital society, for example, by forming public opinions on the net, promoting the IT industry with early adoption of new telecommunication media, and creating a lot of digital cultural content in cyber space. Netizens have become a symbol of the present Korean digital society.

Several media has praised the growth of Korea in IT industry. The New York Times glorified Korea as the realization of "America's broadband dream" (Belson&Richtel, 2003). Paul Wolfowitz who is the former president of the World Bank and a well-known neo-conservative, described South Korea as a "world IT leader" in 2006. The Korean government has succeeded in getting high praise from the outside world, and has persuaded other countries that pervasive broadband, actively subsidized by the government, will increase industrial efficiency, create e-businesses and jobs, improve global competitiveness, and increase household income (Kwang-suk Lee, 2011).

In the information age, people are using the Internet and SNS as a main communication media. They create a public sphere, by the use of Internet and SNS, which are fast, correct, two-way interactive, economic, and digital multi-media. The Internet and SNS have various positive sides. They promote the communication by people of political information and opinion freely and actively. They make people form groups against the wrong-doing of big companies or governments, and keep people's privileges. People buy goods more cheaply via e-commerce, which also helps companies lessen the expenses and heightens the competitive power.

However, the Internet and SNS have several negative sides, too. The government and companies can accumulate people's information and use it for diverse purposes. If the government uses that information to monitor and control people, such a society may become a 'surveillance society' threatening democracy as in George Orwell's '1984'. As companies also use that information to sell products, so people often feel they are captivated by a lot of commercial messages, including much spam-mail. The more society depends on the Internet, the more it can become fragile and risky.

The digital media guarantees neither democracy nor a surveillance society. It's true that new technology gives us the opportunity to expand democracy. However, if we don't use new media properly, we may find it useless or even harmless. Therefore, the positive use of the digital media and keeping our society democratic is very important.

In this paper, I will first study the basic subjects that influence the appearance of the digital society in South Korea. I selected three subjects outside the Internet and SNS to examine: the government (state), the market (capital), and users (citizens). These were continuously examined in my previous analyses to the Korean computer-mediated communication (CMC media) industry (Lee J, 2005; 2011). I will argue for the citizen to promote democracy in the digital society.

Research methods for two research problems are literature review, case analysis, and a political-economic approach. My assertion and theory will be supported by several cases, which were reported as social issues in South Korean media.

2. Research Question and Theoretical Framework

2.1. Research Question and Terminologies

On the basis of this fundamental idea, the following two questions are raised to study the role of citizen and digital media in the knowledge-based society.

First, how can the Korean Internet and SNS (CMC media) industry grow after its birth?

Second, what is the role of three subjects in the digital society?

I will use the following terminology. First, I define "computer-mediated communication (CMC)" as humans' producing and sharing information with others by the use of computers and communication. In other words, CMC means the activity of the Internet and SNS or communication over the Internet and SNS. Its broad meaning includes mobile telecommunications and all other phenomena related to computer-communication. Its narrow meaning is restricted to the cyber space which is comprised of the Internet and SNS.

Second, I define "subject" as a main agent of society influencing the decision of how to initiate, invest, and adopt the Internet and SNS (CMC). Three subjects addressed in this paper are 'the state', 'the market', and 'citizens'. Besides these, other subjects participate in operating and developing CMC. Nonetheless, I will focus on these three main subjects to clarify the interactions of these subjects. The role of digital media is primarily decided by these main subjects.

Third, "digital media" is defined as electronic media that work on digital codes. As to Wikipedia (http://en.wikipedia.org/wiki/Digital_media) digital media like digital audio, digital video and other digital "content" can be created, referred to and distributed via digital information processing machines.

Fourth, "role" is defined as an expected behavior or function in a given society.

Fifth, "digital society" is defined as a society which depends on digital technology and information of which major product and raw material is

2.2. Research Method

I tried to answer the questions through a political-economic approach. The political-economic approach has been regarded as scientific, systematic, and critical research method to capitalistic society. Its basic logic is that mass media has been looked upon as an economical and ideological substance. It helps us to understand the structural peculiarities. It accounts for not only the effects, but also the causes of digital society.

There are also doubts whether it can be applied to the study of information society[1]. It is plausible that a changing society demands a change of research approach. Nevertheless, we cannot delay doing research until that change of approach occurs. Still, we can find or create a new persuasive approach by applying and appraising a political-economic approach to the on-going phenomena of the Internet and SNS (CMC). Garnham (1990) emphasized that the development of a new means of communication is the result of changes in the system of capitalism.

[1] You can see further discussion of the political-economic approach in Vincent Mosco (1989; 1996) and McChesney (1998; 1999).

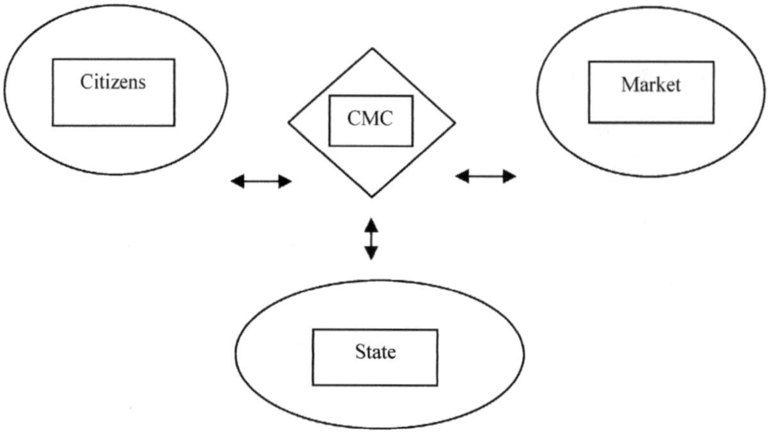

Figure 1. The Structure of Korean CMC (Internet and SNS) Industry*

* State (policy maker) + Market (Capital; Internet service investor) + Citizens (Netizen; Internet users) = CMC (Digital media) and Democracy

The rise and expansion of digital society depends on several elements- the state, the market (e.g. conglomerates, companies, and capital interests), citizens (e.g. netizen, Internet users, and people) and so on. Yve de la Haye (1979) called this phenomenon the dynamic "State-Capital-Social Movements" in which the triangle of forces and their articulation conditions the communication sector. McChesney (1998) explains that the political economy of communication entails two main dimensions; the nature of the relationship of media and communication systems to the broader structure of society, and the influence of ownership, support mechanism (e.g. advertising), and government policies to media behavior and content. Therefore, I will analyze the digital society by focusing its relationship to 'the state', 'the market', and 'citizens'.

2.3. Views of Information and the Internet
The Citizen who are Internet users, form the digital society for better sharing of useful information. Above all, we should look closely at the nature of information and the Internet from several viewpoints. Four viewpoints which are reviewed here are 1) the economical view of information, 2) the e-business view of information, 3) the futurist view of information, 4) the communication view of information and 5) the Panopticon view of information.

First, the economic view of information explains its cost and price, consumption, the role of technology, etc. We can find a technological initiative of information economy from Gordon Moore (1965). He had an insight to key points of information technology. Moore's law of processors is the processing capacity of chips doubles every 18 months. After observing an exponential growth in the number of transistors per integrated circuit, Moore predicted that this trend would continue (Moore, 1965). George Gilder (2000) also supports the powerful influence of information technology. Gilder's law of bandwidth is that available bandwidth doubles every six months. Gilder also said that communications bandwidth is not only the secret of electronic progress, but also the heart of economic growing, and that this will give the masses of people unprecedented hope over industrial revolution (Gilder, 2000)

Shapiro and Varian (1999, 3) said that information is essentially anything that can be digitized-encoded as a stream of bits, and added that information is costly to create and assemble but cheap to reproduce.

"Economists say that production of information good involves high fixed costs but low marginal costs. The cost of producing the first copy of information good may be substantial, but the cost of producing (or reproducing) additional copies is negligible. This sort of cost structure has many important implications. For example, cost based pricing just doesn't work: a 10 or 20 percent markup on unit cost makes no sense when unit cost is zero. You must price your information goods according to consumer value, not according to your production costs."

Information became at once a main source of production and a valuable commodity for consumption. Information consumers react to information producers by copying information. Consumers also value information by experiencing it. "Nowadays the problem is not information access, but information overload" (Shapiro and Varian, 6). So it is very important that users know well how to locate, filter, and communicate what is useful to them. The information economy and society is driven by information technology and "infrastructure that makes it possible to store, search, retrieve, copy, filter, manipulate, view, transmit, and receive information" (Shapiro and Varian, 8). From this viewpoint, we assume that to live in an information society, information plays a key role in economy.

Second, the e-business view of information explains that the Internet, the combination of computer and communication, doubles its usefulness with every new user (Gerbert et al., 2001). This is called Holmes' law[2], and it clarifies why the Internet is a superior collaborative tool. Holmes' law demonstrates that grows much faster than indicated by the commonly quoted 'Metcalf law' for one to one communication (Gerbert et al., 2001).

Metcalfe's law, named after Bob Metcalfe, states that ten-fold increase in the size of a network leads to a hundredfold increase in its value. When there are n people in a network, and the value of the network to each of them is proportional to the number of other users, the total value of the network to all the users is proportional to $n(n-1) = n^2 - n$ (Shapiro and Varian, 1999)[3]. In other words, this law counts the number of one-to-one connections, i.e. the number of subsets with two members. This number is $N(N-1)/2$, or $N^2/2 - N/2$, so it is referred as "growth with N^2" (Gerbert et al., 2001).

However, Holmes' law counts the number of many-to-many connections, i.e. the number of subsets with an arbitrary number (bigger than one) of members. This number is $2^n - (N+1)$, so it is referred as "growth with 2^n". If n is 100, then the value in Holmes' law is about ten to the power of thirty, but the value in Metcalfe's law is ten to the power of four. Therefore Holmes' law produces more value than Metcalfe's law.

These effects arise from 'network externalities'. The network begins from product, but gains power through users. A sufficient size of a network provides the network manager a consecutive market. A network is like a standard. If we adopt a certain standard, then we should buy products related to that standard. Externalities, like feedback, make a network bigger and better, and benefit members more (Shapiro and Varian, 1999).

These rules mean that information technology has the possibility to produce extraordinary value by networking people for the purpose of communicating information. From these laws, the Internet, which is network of networks, creates enormous value because it is the worldwide network itself.

Third, the futurist view of information explains that growing information changes society. Daniel Bell (1973), Alvin Toffler (1980), and others think that a few developed countries have moved from being industrial economies and societies to becoming information economies and societies. After

[2] They (Gerbert et al., 2001, 44) stressed that "Holmes' Law" is not a "law", but merely a convenient shorthand for the tremendous opportunity that lies in making many-to-many interactions work.

[3] Shapiro and Varian(1999, 184) says that if the value of a network to a single user is $1 for each other user on the network, then a network of size 10 has a total value of roughly $100, in contrast, a network of size 100 has a total value of roughly $10,000. Metcalfe told Shapiro and Varian that this law owed George Gilder a lot.

Daniel Bell (1973) focused on how the advanced economies were shifting from being based on the production of goods to a service and information processing basis, others focused more specifically on the central role of information as compared to other services - where wealth and economic activity are created, what grows most, and where most people are employed (Straubhaar & LaRose, 1997, 58).

Daniel Bell (1973) says that manipulation of information makes us a "post- industrial" society, an information society. In an information society, more people are employed in various information media, and communication industries. Futurists, such as Bell (1973), think computers and information technology contribute to economic development, thus extending productivity growth. They also see that technology will provide the effective instruments to manage social conflict in a manner not possible under industrial capitalism (Bell, 1973)

Computer communications systems are expected to advance economy and politics by allowing higher productivity and widespread political participation (Toffler, 1980). As the information technology develops, the source of power is also shifted from money [capital] to information [knowledge] (Toffler, 1991).

Futurists are optimistic about the development of technology, because they see it transforming society through the advancement of information society. They believe that computers and communication advance economic welfare and political democracy, which are limited to relatively fewer, powerful elites in capitalist society.

Fourth, the communication view of information explains that the Internet, a global network of computer networks, is an emerging mass media, with especially diverse expression modes. Anyone who is connected on-line is able to send e-mail, chat with others at the same time, and search for data anytime, anywhere. The Internet and SNS (CMC) as a new media are distinguished from the old media by five qualities; packet-switching, multimedia, interactivity, synchronicity, and hypertextuality (Newhagen & Rafaeli, 1996).

These five unique qualities make the Internet and SNS more competitive new media. First, packet-switching makes transmission and reception of mass information technologically safe and fast. It sends messages via packets of information and receives them by reassembling them for safe and fast transmission. Second, multimedia empowers the Internet to present rich contents. The Internet and SNS (CMC) facilitate diverse forms of text, sound, images and animation. Third, interactivity attracts more users to the Internet, because it meets the user's presenting desire. Feedback by interaction makes the user of the Internet not a passive object but an active subject. Fourth, synchronicity is one of main reasons Internet users feel the liveliness of communication. When users interact with partners in real time with Internet Relay Chat (IRC[4]), their synchronistic feedback overcomes the barrier of distance. It does not exclude asynchronicity when their exchange of messages involves lag time. Fifth and finally, hypertextuality gives us customized information. Users have more ability to choose information that they want to acquire. This nonsequential message processing allows both traditional linear reading and multi-linear reading. These five characteristics of CMC make the Internet superior to other traditional modes of communication.

Five common forms of CMC are electronic mail (e-mail), bulletin board system (BBS), IRC, multi-user domains (MUDs), and the World Wide Web (Wood and Smith, 2001). The Internet and SNS also provide a virtual reality, a technology that gives users a realistic, three-dimensional, interactive experience. Virtual means "not in fact"; reality means "in fact", and virtual reality then means not-in-fact fact (Dominick 1996). Therefore the Internet and SNS which are convenient, cheap, effective, and interactive media, become the dominant media over newspapers or television.

This view of communication researchers tries to understand the inherent functions of the Internet and SNS, and to apply them to social life for a better society. They also believe that new media may

[4] IRC is like a real conversation. Chatting room and messenger service make participators to send and receive opinions and information in response to their feelings, judgments, and positions. It substitutes real conversation by the advantages of less expenditure, faster facing, and overcoming far distance.

lead to a new society. McLuhan (1966) distinguished societies, and welcomed the electronic age with television before the Internet began to appear in 1969.

Fifth, the Panopticon view of information explains that the Internet and information system can be used as a tool of controlling people. This concept was designed as a model prison by Jeremy Bentham (1787). The Panopticon which means that all can see, functioned as a round-the-clock surveillance system. The design of Panopticon made it possible that no prisoner could ever see the 'inspector' who conducted surveillance from the privileged central location within the radial configuration. The prisoner could never know when he was being surveilled. Therefore, such system with prison's mental uncertainty could be a crucial instrument of discipline.

Michel Foucault (1975, 195-228) developed this concept for described the implications of surveillance society.

"Hence the major effect of the Panopticon: to induce in the inmate a state of conscious and permanent visibility that assures the automatic functioning of power. So to arrange things that the surveillance is permanent in its effects, even if it is discontinuous in its action; that the perfection of power should tend to render its actual exercise unnecessary; that this architectural apparatus should be a machine for creating and sustaining a power relation independent of the person who exercises it; in short, that the inmates should be caught up in a power situation of which they are themselves the bearers. To achieve this, it is at once too much and too little that the prisoner should be constantly observed by an inspector: too little, for what matters is that he knows himself to be observed; too much, because he has no need in fact of being so. In view of this, Bentham laid down the principle that power should be visible and unverifiable. Visible: the inmate will constantly have before his eyes the tall outline of the central tower from which he is spied upon. Unverifiable: the inmate must never know whether he is being looked at any one moment; but he must be sure that he may always be so. In order to make the presence or absence of the inspector unverifiable, so that the prisoners, in their cells, cannot even see a shadow, Bentham envisaged not only venetian blinds on the windows of the central observation hall, but, on the inside, partitions that intersected the hall at right angles and, in order to pass from one quarter to the other, not doors but zigzag openings; for the slightest noise, a gleam of light, a brightness in a half-opened door would betray the presence of the guardian. The Panopticon is a machine for dissociating the see/being seen dyad: in the peripheric ring, one is totally seen, without ever seeing; in the central tower, one sees everything without ever being seen."

Panopticon is no more a concept in the digital information age. Considering it as a big data system of monitoring and documentation in database, we live in the Panopticon society.

3. The Subjects of CMC Development and the digital society in Korea

3.1. History of the Korean CMC Industry[5] as the Digital Media Industry

Since the establishment of the first Korean CMC Company, Dacom, in 1982, several other companies, including Korean PC Communication (1992), Nowcom (1994) and Samsung SDS (1996), have followed. In this early phase, the service type of CMC was different from that of the Internet because of the underdevelopment of information technology. In spite of the short history of the Korean CMC industry, it expanded rapidly with CMC service users, amounting to 578 thousand people in 1994. By the way, in that same year, Internet services were officially introduced by the Kornet of Korea Telecommunication (KT). In the beginning of Internet services in Korea, only a limited number of domestic CMC users could contact the Internet. Then, as domestic CMC services allowed their users to connect freely to the Internet, the number of Internet users became similar to that of the CMC. However, after most of the domestic CMC users connected to the Internet, the relationship between domestic CMC and Internet use reversed. The increase of Internet users didn't depend upon domestic CMC companies anymore. People were able to use Internet services without subscription to domestic CMC service. Two portal services, Naver and Daum , have been survived after severe competition

[5] This analysis was based on the writer's thesis (Lee J., 1997; 1999, 2005) and partly revised.

since 1998. They succeeded to gather a lot of subscribers and launched SNS later. People chose the Internet instead of domestic primitive CMC service. Domestic CMC services began to show slower growth, to recede and be transformed into portal sites of the Internet. Therefore, CMC service is now equivalent to the Internet and SNS.

The developing trend shown in Table 1, Table 2, and Table 3 illustrates that CMC became one of the major communication phenomena in Korea. The number of smartphone users in Korea has exceeded 20 million on November 1st 2011. In other words four out of 10 Koreans and eight out of 10 economically active persons in Korea are currently using one of these electronic gadgets. It took just 2 years since the first smartphone first introduced in the Korean market in November 2009. The recent data shows that smartphone users are 60,559 thousand in August, 2016 more than all the population.

Table 1. Increase of Internet Users in South Korea

Year	1994	1996	1998	2000	2004	2010	2015
Users*	138	731	3,103	19,040	30,670	37,010	41,940

* The unit of users over three years old is one thousand persons. Source: National Computerization Agency (NCA), 2004-2010 and Korea Internet and Security Agency (KISA, http://isis.kisa.or.kr/), 2015; The rate of Korean Internet users in 2015 is 85.1%.

Table 2. High Speed Internet Subscribers in South Korea

Year	1999	2000	2001	2004	2010	2015
Subscribers	266	3,939	7.763	11,921	17,219	20,400
Family Ratio	1.9	27.4	52.3	76.6	100.4	97.1

* The unit of users is one thousand households. Source: MIC, * Estimated total households of Korea for 2004, 2011, and 2015 are 14,311, 17,152, and 21,011 thousand. Family ratio is the rate of family using high speed Internet service including cable modem.

Table 3. SNS usage rate over 6 years old in South Korea

Year	2010	2011	2012	2013	2014	2015
Ratio	67.1	66.5	67.1	55.1	60.7	64.9

* Out of the Internet users ages 6+, 64.9% have used SNS (Social Networking Service) in the last 1 year, while 46.1% access SNS 'In the last 1 week'. SNS strengthens personal relationships with acquaintances or creates new social networks to broaden one's network of relationships, includes Minihompy (small homepage), Blog, Twitter, Kakao Story, Facebook, and Google+. Source: KISA, http://isis.kisa.or.kr/

To examine the economic characteristics of the Korean CMC Industry, this study divides its history from 1982 to 2016 into four periods on the basis of the telecommunication policy of the government, the competition situation of the market, and citizens' using CMC services. I also tried to identify what subjects influenced the development of CMC. The first period is the formation stage or monopoly period (1982-1990). The second is the growth stage or oligopoly period (1990-1995), the third is the prosperity stage or domestic competition period (1995-2010), and the fourth is shift stage or global competition period (2011-2016).

In the first stage, the state played a large role by initiating the foundation of Dacom, supporting its management by creating a large government demand, and deregulating data communication institutions. President Chun Doowhan (1980-1988) agreed to this plan which a computer manager proposed in 1980 for economic accomplishment. A public enterprise, Korea Telecommunication which held 33.4% of the shares, and various private enterprises jointly founded the Dacom Company. Dacom was the first company authorized as a value added network (VAN)[6] enterprise dealing with data communication.

There were four deregulation policies implemented in the 1980's. The first deregulation, in 1985, introduced the approval system, where a person or corporation could offer data search and treatment services. The second deregulation, in 1987, expanded services from the company subscribers to the leased lines for general use. The third deregulation, in 1988, permitted medium and small-scale companies to build computer networks and national infra computer network enterprises. Lastly, the fourth deregulation, in 1989, expanded the scope of the multiplexes.

The government intended to promote industrial development by enhancing the Korean CMC Industry, which was predicted to greatly affect the electronic industry and other industries. To this purpose, the government also tried to increase the effectiveness of the administration by computerization and implementation of a network system.[7] When the government becomes stream-lined and effective, then it possibly gives hegemony (leadership) to the market (capital) and citizens.

In the second stage, the market was a major subject in that Korean conglomerates actively participated in the CMC service market to achieve a rapid expansion of capital and make a good profit. As the government transformed the monopolistic market with Dacom to a competitive market with multiple enterprises because many advanced countries were expected to invest on a large scale in the information and communication industry, the public enterprise Korea Telecommunication and several conglomerates companies were concerned with the various effects of participation in the CMC industry.

In the third stage, citizens also became a main subject by using Internet services actively. After the 'notice policy' of the government in 1995, venture capital increased, investing a lot of Internet venture companies in free communication market. A more competitive communication market made citizens as consumers more powerful. Citizens had more opportunities and means to address the wrong doings of the state and the market. Citizens' subscribing to mobile telecommunication, one of broader CMCs, was also rapidly increasing. Also, citizens armed with two CMCs, nourished political power to affect the election of parliamentary members in 2000, and the president of the Korean government in 2002.

The citizen's rise and the capital's enduring power do not imply the state's weakening. The state also succeeds to keep its role, developing CMC industry and educating computer illiterate. The accomplishments of CMC policy verify effectiveness of government administration. The information and communication minister Daeje Chin (2004) emphasized an advance in information technology (IT) field including CMC.

"As of the end of 2003, 11.18 million households – more than 73% of the total number of households – subscribed to broadband Internet and 29.22 million people – 66% of total population – had access to the Internet. According to the 'ITU 2003 Internet Reporter,' Korea ranks first in terms of broadband Internet penetration rate, has the third largest population of Internet users, and has the fourth highest PC penetration rate in the world. These statistics will illustrate Korea's firm standing as a leading Internet country."

---

[6] VAN (Value added network) enterprises in Korea are called enhanced services in the U.S.A. and VADS (Value added data service) in the U.K.

[7] The subsequent government has reinforced this view. The following is offered from the ministry of information and communication (MIC).

"Informatization" is the core national strategy for such environments and it is an important tool for removing social inefficiencies and enhancing productivity and transparency of a society. The IT industry, the supply base for informatization, leads the growth of a knowledge-based economy through creating value-added, new business opportunities and jobs. (http://www.mic.go.kr/english/, 1999)

Telecommunication industry, CMC industry related, of Korea posted the biggest trade surplus, $13 billion in 2003 among member countries of the OECD[8]. Korea also exported $154.0 billion worth of IT goods and posted an IT-related trade surplus of $78.2 billion in 2010. These statistics are compliments to both the government and the market. The conglomerates including Samsung Electronics co. and LG Electronics Inc., got enormous profits and reinforce their power to the government and citizens.

In the fourth stage, the CMC industry has some difficulty in the field of both hard ware and soft ware. In the hardware, smartphone became a main media for the use of the Internet and SNS. Samsung Electronics has achieved a very large profit in the smartphone market. But the future promise is not bright, because technical problems of hard ware may weaken the brad power. In the soft ware, as smartphone adopted new types of services such as social media, SNS services expanded their market. For the CMC industry to keep growing continuously, workers and users need free and creative working condition. However, bureaucratic, less democratic government and short-term profit seeking big companies in South Korea demanded CMC workers follow managers' guidelines for the past ten years (2008-2017). These strict organizational environment prevented interactive communication. Less communication leads bigger fails by smaller mistakes. If any country, including South Korea, would not advance democracy and improve two-way, interactive, open, equal communication, further development of CMC industry cannot be expected any more.

3.2. The role of CMC industry as Digital Media

With CMC, since the 1990s there have been three characteristics: 'information hegemony' of the state, 'information economy' of the market (capital) and the 'information democracy' of citizens in Korea. These findings give us a new evaluation of the roles of Korean CMC industry.

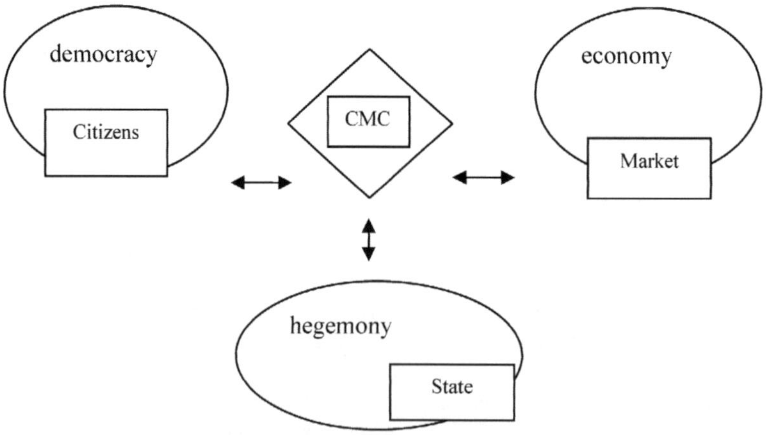

Figure 2. The Characteristics of Korean CMC Industry*

* Korean CMC industry's characteristics = State (information hegemony) + Capital (information economy) + Citizens (information democracy)

First, each different subject that influences the development of the Korean CMC industry at each phase shows that the social adoption of CMC in Korea is dynamic and changeable according to the social conditions. Secondly, three interests - the state's information hegemony, capital's information economy and a citizens' information democracy - access CMC actively each out of its own necessity. In addition, these characteristics of Korean CMC industry are also changeable due to cooperating and

---

[8] In 2003, South Korea's surplus in telecommunication equipment was $13 billion, followed by Mexico's $11 billion and Finland's $7.2 billion, Japan's $6.6 billion, and Britain's $5.5 billion. www.korea.net

conflicting relationships between these subjects. Lastly, the subject that the roles of each subject have two sides, both positive and negative, means that a more prudent, harmonious, and democratic Internet and SNS policy is needed.

Shortly, Korean CMC industry has grown a lot in the 1990s. Three points are drawn as main causes of this growth. First, the government had a political consensus to a development of CMC and fully supported it. The government regarded CMC as a key leading industry and encouraged the market to invest it. Second, the conglomerates such as big company groups searched for a promising industry to create demand largely, and they saw CMC as a good alternative. In particular, they had large infrastructures in IT industry, so they could launch and expand CMC industry easily. Third, many new creative researchers join the IT industry, and they contributed the growth of CMC industry. Lastly and fourth, citizens eagerly need new media to serve their own interests, and they found it not in traditional newspapers or televisions, but in CMC. We should observe that the 1990s of Korea is transition period from authoritative underdeveloped society to democratic developing society. Such period is on the one hand unstable, on the other hand dynamic. Citizens need more information about political and economic situations. Furthermore, the government toward democratization, gave citizens more freedom of expression, and allowed CMC to flow diverse information. Therefore, these four causes made Korean CMC industry rapidly growing.

4. The Role of Citizen in the Digital Society

The digital society appeared and expanded by needs of three subjects to the digital media by the help of information and communication technology (ICT). The state tried to administrate effectively by using the digital media actively. The digital media made it easy for the government to accumulate peoples' private information en masse and to manage them. Such society is fragile to the threat of wrong leaders who want to be authoritarians. But it is very difficult to draw a fine line between effective administration and privacy deprivation.

The market tried to get more profit by using the digital media actively, too. The digital media made it easy for the market to accumulate consumers' private information en masse and to sell them goods and services. In that society citizens are often treated as not subjects or objects of economy but means for capital's profit. If the market weighs only profit, then the digital media is no more than a surveillance tool captivating citizens.

Citizens also tried to get more information by using digital media actively. The digital media made it easy for citizens to keep the government clean and productive. If the market sells bad products, then they can form a group, demonstrate or demand to correct its faults. But citizens' resistance is relatively restricted because it depends on the condition that the level of surveillance by the state and the market to citizens is low.

Korea may be called the digital society in the information age that most population uses the Internet and SNS via broadband or mobile phone[9]. It is not unusual that the Internet and SNS often appear as the social issues.

5. Conclusion and Implication

Korea is a very interesting case. Korea shows rapid growth of the Internet users, and larger trade surplus in telecommunication industry with the help of government's successful information technology policy. And Koreans also made their country more democratic with active participation.

This paper analyzed the growth of the Internet and SNS in South Korea. The Internet and SNS created cyber space as a public sphere. They have several advantages as an effective means of communication. Cyber space is influenced by three subjects such as the government [state], the market [capital], and citizens [people].

---

[9] Information society can be differently named by several researchers; 'network society' (Jan Van Dijk, 1999), 'pay-per society' (Vincent Mosco, 1989) etc.

There are two research questions. First question is how the Korean CMC industry can grow fast after its birth. Three main subjects were dealt with in this research. They are the State, the Market, and the Citizen. I divided the history of Korean CMC industry into three periods. The first formation stage is from the birth of CMC in Korea between from 1980 and 1990. The government initiated the monopolistic CMC market. Several conglomerates participated to co-operate the government. But the users are very small. The second growth stage is between from 1990 to 1995. The government also deregulated the Market with changing policy from 'appointment' to 'registration'. The companies increased investment for the possibility of wide diffusion of CMC use. The third prosperity stage is between 1995 and 2010. The government promoted the CMC market's competition with 'notice' policy. And citizens actively enjoy and apply CMC services. However, the fourth shift stage to smartphone faced several problems such as less democracy and one way communication which will weaken the creativity of the content.

Second question is what the roles of three subjects are. I examined the cyber space by the uses of digital media with three subjects. Even though the state and the market have limits to promote democracy, the citizens are expected to make the digital society more democratic. If the state tries to monitor citizen by the use of digital media, that society is called 'surveillance society'. As the government is open and transparent, the citizens' democracy will increase. The market should limit to gather and accumulate peoples' information and profile for protecting their privacy.

Cyber space is two-way, economic, and open to every people. It has various positive sides. It promotes the communication by people of political information and opinion freely and actively. It makes people form groups against the wrong-doing of big companies and keeps people's privileges. People buy goods more cheaply via e-commerce, which also helps companies lessen the expenses.

However, the cyber space has several negative sides, too. The government and companies can accumulate people's information and use it for diverse purposes. If the government uses that information to monitor and control people, such a society may become a 'surveillance society' threatening democracy as in George Orwell's '1984'. As companies also take advantage of big data to sell more products for their profit increase, so people often feel they are captivated by a lot of commercial messages, including much advertising spam-mail. The more civil society depends on the cyber space, the more it can become fragile and risky.

The digital media guarantees neither democracy nor a surveillance society. It's true that new technology gives us the opportunity to expand democracy. However, if we don't use new media rightly and positively, we may find it negative or even harmless. Therefore, the positive use of the digital media and keeping our cyber space democratic is very important.With democracy, the Internet and SNS industry can be flourished by the cooperation of netizen, business managers, IT researchers, and politicians.

References

[1] Baym N K 1997 Interpreting soap operas and cheating community: Inside an electronic fan culture *Culture of the Internet* ed. S Kiesler (NJ: Lawrence Erlbaum Associates) pp 103-20.

[2] Bell D 1973 *The coming of post industrial society* (New York: Basic Books).

[3] Belson K and Richtel M May 3rd, 2003 America's Broadband Dream Is Alive in Korea *The New York Times*.

[4] Bentham J 1995 Panopticon *The Panopticon Writings* ed. M Bozovi (London: Verso) pp 29-95.

[5] Borenstein D 2000 *Quoteland.com*. <http://www.quoteland.com>

[6] Cohen H 1999 Economics and the Internet: No free rides on the InfoBahn the information superhighway becomes a toll road *Real law @ virtual space* ed. S Drucker and G Gumpert (Hampton press) pp 95-126.

[7] December J 1997 John December What is Computer-Mediated Communication... *http://www.december.com/john/study/cmc/what.html*

[8] Dijk J V 1999 *The network society* (Sage).

[9] Dominick J R 1996 *The Dynamics of Mass Communication* Fifth Edition (McGraw-Hill).
[10] Dyer-Witherford N 1999 *Cyber-Marx: Cycles and Circuits of Struggle in High Technology Capitalism* (Urbana and Chicago: University of Illinois Press).
[11] Foucault M 1995 *Discipline and Punish: The Birth of the Prison* (translated) A Sheridan (NY: Vintage Books).
[12] Foucault M 1980 *Power/Knowledge: Selected Interviews and Other Writings 1972-1977* (New York: Pantheon).
[13] Gerbert P, Birch A, Schnetkamp G and Schneider D 2001 *Digital storm* (Oxford: Capstone).
[14] Gilder G 2000 *Telecosm: How Infinite Bandwidth will Revolutionize our World* (Free Press).
[15] Hague B and Loader B 1999 *Digital democracy: discourse and decision making in the information age* (Routledge).
[16] Hirschkop K 1998 Democracy and the new technologies *Capitalism and the information age* ed McChesney *et al* (Monthly review press).
[17] Korea Information Society Development Institute (KISDI) 2003 *IT industry outlook of Korea.*
[18] Lee J 1997 *A study on the formation and growing process of the Korean computer-mediated communication industry* PhD thesis (Seoul: Kyunghee University).
[19] Lee J 1999 *Ideology in Information Society* (Seoul: Communication Books).
[20] Lee J 2001 The Change of the Korean Mass Media and its Ideological Role since 1980: from the Newspaper and its Ruling Ideology to the Computer Mediated Communication and its Balanced Ideology *Korean J. of Jour. & Comm. Std.* **46-1** pp 1-16.
[21] Lee J April 2nd-3rd,2005 Internet and Society in Information Age: Possibilities and Limits in Korea *Paper for presentation at the 2005 Magness Inst. Acad. Sem. at NCTA in San Francisco.*
[22] Lee J September 27th-29th,2006 The Role of Citizen and Digital Media for Democracy in the Knowledge Society *Paper for presentation at the Int. Conf. on The Media and Democracy in the Knowledge Society at Zelkova Hotel in Seoul.*
[23] Lee J November 25th-26th,2011 New Media and the Knowledge - Based Society: Possibilities and Limits in Korea *Keynote Speech Paper at the Int. Conf. of UMN in Serpong-Indonesia.*
[24] Lee K S October 28th,2011 Interrogating a Janus-faced "Digital Korea": To Restore Democratic Values of Information and Communication Technologiesat the Local Scene" *The Int. Sem. on Future Visions and Suggestions for the Next Generations of Journalism and Mass Communication Studies: Rethinking Media and Democracy* (Sungkyunkwan University).
[25] McChesney R W 1998 The Political Economy of Global Communication *Capitalism and the Information Age: The Political Economy of the Global Communication Revolution* ed R W McChesney *et al* (New York: Monthly Review Press) pp 1-26.
[26] McChesney R W 1999 *Rich Media Poor Democracy: Communication Politics in Dubious Times* (Urbana and Chicago: University of Illinois Press).
[27] McLuhan M 1964 *Understanding Media: The Extensions of Man* (New York: McGraw-Hill).
[28] Moore G E 1965 Cramming More Components Onto Integrated Circuits *Elect.* **38-8**.
[29] Mosco V 1989 *The Pay-per Society* (New Jersey: Ablex Publishing Corporation).
[30] Mosco V 1996 *The Political Economy of Communication* (Sage).
[31] National Communication Agency (NCA) and Ministry of Information and Commmunication (MIC) 2004 *White Paper Internet Korea 2004.*
[32] Newhagen J E and Rafeli S March 1996 Why communication researchers should study the internet: A dialogue *J. of Comp.-Mediated Comm.* **1(4)** http://www.ascusc.org/jcmc/vol1/issue4/rafeli.html.
[33] Rosen L D, Deborah C S and Michelle M W 1987 Computerphobia *Behavior Research Methods, Instruments, & Comp.* **19** pp 167-79.
[34] Scott C R and Steven C R 1997 The effect of communication, writing, and technology apprehension on likelihood to use new communication technologies *Comm. Educ.* **46** pp 29-43.

[35] Shapiro C and Hal V 1999 *Information Rules* (Boston, MA: Harvard Business School Press).
[36] Straubhaar J and Robert L 1997 *Communications media in the information society* (Wadsworth).
[37] Toffler A 1980) *The third wave* (New York: Morrow).
[38] Toffler A 1991 *Powershift: Knowledge, Wealth, and Power at the Edge of the 21st Century* (Bantam Books).
[39] Whitaker R 1999 *The End of Privacy* (New York: The New Press).
[40] Wood A F and Matthew J S 2001 *On line Communication: Linking Technology, Identity , and Culture* (NJ: Lawrence Erlbaum Associates).
[41] Yve de la H 1979 The Genesis of the Communication in France *Communication and Class Struggle* ed A Mattelart and S Sieglaub (New York: International General) **1** pp 201-5.

Good practices in Local Government - A first overview of Portuguese reality

P. Carvalhosa, F. Portela*, J. Machado, M.F. Santos, A. Abelha

Algoritmi Research Centre, Guimarães, Portugal

a64868@alumni.uminho.pt; {cfp,mfs}@dsi.uminho.pt; {jmac,abelha}@di.uminho.pt

Abstract: Good practices in eGov are being increasingly used by Local Governments being that it is considered by them as an advantage. The main goal is providing to the town hall a differentiation point and approximate their services to the citizens. For this, it is necessary to define and apply innovative strategies in order to increase the use of services by the citizens. This paper is framed in a research work and it presents a first overview of the existing good practices in eGov, taking in consideration the Portuguese's reality. The good practices identified were distinguished with many awards and with a positive response from the target audience. The use of digital marketing strategies aims to increase their membership and coming closer the municipalities of its citizens through the dissemination of the good practices. At this moment the data collected are almost exclusively of good practice in Portugal, however some international practices were also identified. As a result of this study the community has a list of good practices that can be applied in their municipalities.

Index Terms: eGov, Local Government, Good practices, Town hall, Municipalities, Digital Marketing

1. Introduction

The municipalities increasingly need to adapt their services to provide better services to the citizens. The town hall must balance the need to transmit information and capture information, based on a set of factors, such as population, topographic, climate and economy. Local governments must define their own goals and progress reports in accordance to the interests and expectations of the citizens. So, each town hall needs to determine the objectives and the appropriate level of citizen participation[1]

The town hall has a permanent management function of municipal services and activities. The town hall creates local laws and obligations to citizens. These laws are reviewed, approved and after that are provided by the services, in the municipality[2]. The services included are: Energy; Transport and Communications; Education and Culture; Leisure spaces; Health, social action and housing; Civil Protection; Environment and Sanitation; Protecting the rights of consumers; Land-use planning and Urbanization; External Cooperation; Urban police. The town halls need to be able to respond quickly and easily perform all the tasks. Consequently, municipalities have adopted the use of electronic Government (eGov) services. The eGov is a generic term to refer governmental services which are based on the Web. The government uses information technology and particularly the Internet to support their operations, engage citizens and provide public services. The interaction can be used to obtain information, filings, or making payments and a number of other activities through the World Wide Web. In summary, it aims to promote a more effective government, make the government services more accessible, allow greater public access to information and makes the government more accountable to the citizens. eGov involves services Internet, telephone, community centres, wireless devices and other communication systems[3,4].

Local eGov is a form of governance, which focuses specifically on the services provided by local authorities. The local municipality however is closer to the citizen, for example, the territorial proximity to the citizen is a differentiation point between the eGov and local eGov[2].The main goal of eGov is improving the performance of the organization and the citizen's satisfaction. For that it is necessary doing market studies to obtain a vision about the citizen's needs. So, the eGov combined with a good marketing strategy allows the municipalities to have a privileged interaction with their citizens. This article aims to inform those of what some of the eGov best practices are. The main focus is the Portuguese reality and the benefits of adopting eGov practices.

This paper describes the first results gatheredfrom a research project. This research project has as main goal identifying a set of good practices used at a Local government level. It is important to note that all of the good practices identified need to be focused on the citizen, in their health, security and in the capability to improve the relationship between them and the town hall.

There are studies which discuss good practices that focused on public administration. Studies such as Missouri Municipal League INNOVATION AWARDS [5] and the Good practices and innovations in public governance [6] demonstrates a set of good practices which won innovation and transparency awards. However, these studies only describe the project but did not describe how well the good practices implementation was doing. In order to search the good practices,a benchmarking methodologyis used to define keywords and was done with an annual granularity time. The timeline for this was placed only in the XXI century, where once in the timeline a large technological development was found particularly in the use of Internet.A brief background along with related work is presented in section 2. Section 3 presents the Local eGov. Section 4 briefly describes the Good practices using eGov. Section 5 addresses an introspective discussion of this work. Section 6 presents the Conclusion and Future Work.

2. Background

2.1. electronic Government (eGov)

Local eGov is a level of formal state institutions that provide services to citizens and local organizations, acting in accordance with existing national laws. The electronic government allows local governments another way to serve the citizens and bring innovation to its internal operations. According to Cook[7] the practice of local eGov is reflected in four dimensions: eServices is the electronic delivery of government information, programs and services, mainly of the internet; eManagement is the use of information technology to improve government management. This service enables business processes to be simplified in order to improve the flow of information through government departments; eDemocracy is the use of electronic media such as email and Internet that increase citizen participation in the decision-making process; eCommerce is the exchange of money for goods and services over the Internet, which may include citizens who pay taxes and utility bills, vehicle registrations, and more.

The increase amount of information leads to an increase of the quality in the decision-making process. The organization needs to adopt tools enabling the acquisition, processing and presentation of information for management[8]. The use of BI in organizations is revealed as an advantage in the decision-making process, allowing the differentiation of services in organizations [9]. However, the number of solutions to local government is reduced[8].The government needs to respond to the request for change in order to make public services more effective and efficient. Simplex, a program developed in Portugal aims to simplify services, bring the services closer to citizens, services tailored to the citizens, and do more with fewer resources[9]. Most of these goals can be achieved by the use of Business Intelligence.

2.2. Digital Marketing

Digital Marketing leads to the creation of the product / service using the internet as a key to satisfy the requirements for new forms of innovation. The internet is used as tool to exchange value. By using the internet, products and services are positioned and promoted, purchased, shipped to distribution and requests are made for support and maintenance. The website offers a greater choice, more influence in decisions and more power to the consumers. Brands have new ways to create and sell their products to business services and markets, which results in reaching people more easily.

Digital Marketing is based on understanding how people are using the technology, and how the organization can take advantage of technologies to reach consumers more effectively. The organization works in order to understand what consumers need in order to deliver their products and services, successfully to their customers.Over the years, the use of technology has greatly increased, so the emergence of Digital Marketing was a predictable consequences [10].

In the world, 45% of the population uses the internet. North America (87.9%), Europe (73.5%) and Oceania (72.9%) are the geographical areas where the percentage of population using the internet is more pronounced. On the other hand, Africa (27%) and Asia (38.8%) are the geographical areas where the percentage of population using the Internet is lower [11].Portugal had 7 015 519 Internet users, which corresponds to 67.5% of citizens. One of the digital marketing channels that most usesduring this time was Facebook. Portugal has 5 600 000, which corresponds to 54% of penetration rate.

3. Local eGov

3.1. eGov in Portugal

The 2011 study, of *Agência para a sociedade do conhecimento*[12], made a statistical analysis of the various municipalityservices. This study also had the participation of all the town halls where all the municipalities along with the autonomous regions answered questionnaires.In terms of the results, the study showed:

- In 2011, 68% of Municipalities had an ICT development; All the municipalities had Internet connection; Although all town halls have internet connection, only 37% of municipality employees used the Internet regularly;
- By type of open source software used: 73% Open source software for operating systems, 68% Open source software for Internet servers and 48% Open source software for other type of applications;
- Municipalities, by security applications used: 99% Anti-virus software, 95% Firewall, 91% Anti-spam filters, 53% Secure servers (e.g. https protocol), 40% Backup of information in a location external to the organization. However, security problems were detected: 10% Virus attack without loss of information, 3% Virus attack that caused loss of information, 3% Blackmail or threats to software or data and 2% Non-authorized access to the computer network or data; Municipalities by difficulties and problems experienced in information and communication technologies (ICT) implementation are: 89% Financial resources shortage, 58% Resistance to change, 58% Lack of informatics technicians, 56% Lack of training programs and 27% Management problems;
- Municipalities by activities performed on the Internet: 99% Search and gathering of information and documentation, 98% E-mail and Electronic exchange of files, 92% Promotion of goods and services provided by the Municipality, 91% Communication with other Public Administration organizations, 88% Purchase catalogue consultation, 87% Data bases access and 82% Communication with citizens;
- Municipalities with web presence, by services or functionalities offered in the website have: 94% Forms for download; 82% Public consultations; 80% E-mail to receive messages and information requests, 70% Newsletters subscription, 62% Geographic Information Systems / interactive maps, 44% Reception of garbage collection or street cleaning requests, 43% Online library services (e.g. search / availability, etc.), 39% Employment opportunities, 39% Forms to fill up and submit

online, 38% User support (Helpdesk, FAQ's, etc.) and 35% Make surveys to the citizens; On the other hand, only 7% had online payments, 16% ability to guarantee secure transactions and only 18% Online Service to the public (e.g. Virtual service desk);

- Municipalities by types of concerns considered in the municipal ICT strategy: 67% ICT infrastructure, 64% Online services for citizens, 63% Security policy regarding ICT, 43% Training or certification of ICT competencies and 31% Ordering goods or services through the Internet (e-Commerce).

After the analysis of this data it was possible to conclude that all town halls have internet, but only 68% of Municipalities have an ICT development and only 35% of the town halls conduct surveys on the citizens. Other challenges for the municipalities are the number of employees who regularly use the internet (37%) and the number of municipalities which allow citizens to make online payments (7%). Through the analysis of this data, it is a possible conclusion that there are many ways to become more to the eGov services and to increase the use of their services by the citizens. In conclusion,it is necessary to improve the online payment services and online interaction with citizen's services.

With regard to Europe, Portugal according to EC [13] is one of the most evolved countries in the European Union as it regards the implementation of eGov in public administration. Therefore, Portugal is a great example in regards to their provision of online public services.Portugal also stands out from allthe other countries in transparency and facilitators of information technologies (usability, easy to use and speed of use). In these cases, the transparency presents a percentage of 71% and the facilitators presents a percentage of 83%, the values in Portugal are higher than the average value of the European Union (46% and 68%, respectively).

3.2. Use of Digital Marketing in eGov

The local government does not take much account of the marketing strategies to get the information and services for their citizens. However, the advance of technology in the twenty-first century leads to increase the number of municipalities which have the need to create marketing strategies to reach out to their citizens.The marketing strategy should allow the customer satisfaction measurement in the services provided as well as allow them to make studies of customer satisfaction. This will allow them to provide suitable services. The use of technologies simplifies the search process and allows an easy access to the information found facilitating the citizens' life. The phone or Internet surveys made by the municipality provides a more convenient and easier way to get information [14]

The Town Hall must analyze the Digital Marketing strategies to get closer to their citizens. They can do this by using Digital Marketing to disclose of important events that occurs in the city. Furthermore, for the citizens participating in the decisions-making process, need a well-defined marketing strategy to achieve the objectives of the organization in order toeffectivelyparticipate in the decision making process.

3.3. Benefits of eGov

The eGov goal is to have more interaction between citizens and the municipality. So, it can intensify the demand and consumers choice, increase local competition, reduce the cost of the services provided, improve and innovate the way the local governance is provided to their citizens[15,16].

The main objectives that lead to adopt a strategy to a digital level are the needs of process and documents dematerialisation, reducing the budgets and simplifying the demand-response processes. The electronic government ensures another way to serve the citizens and bring innovation to its internal operations in the local governments. The use of information technology can provide alternative services, facilitate interaction with citizens and the government itself[7,15].

4. Good practices using eGov services

In the world there are several good practices at the level of local eGovernment. Below there are some examples of success, which are briefly explained. The description presents the objectives of services and creative ideas that are implemented by the town halls, with more emphasis on Portugal.

4.1. Best Digital City – Guimarães

In 2015, the Guimarães town hall [17] (CMG) won the Best Digital City award attributed by Agência para a ModernizaçãoAdministrativa (AMA). The criteria as design, customization, innovation and usability in the CMG, through its services and multiple platforms referred to in ICT. CMG was recognized as the best digital city in Portugal. The CMG won the award thanks to their website, its digital services, including the "online services" and "clarify online", the website of the executive, the assembly, the platform for education and the application for Mobitur mobile and the ePaper urbanism platform. These projects and application deployed in Guimarães were appreciated by the association of companies and they are leading the revolution of the Digital Economy in Portugal. Particularly, online shopping, financial websites, portals, media sites, telecommunications companies, e-commerce services and digital marketing.

At this time they are introducing BI technologies in the organization in order to improve decision-making and bring decisions closer to the citizens [8,9].

4.2. PlatformBizfeira – Santa Maria da FeiraandTalk with the president – Setúbal

Santa Maria da Feira[18] aims to bring businesses and citizens around the world who wish to develop contacts and business with companies and citizens from Santa Maria da Feira. The platform Bizfeira is a virtual space for active and collaborative business where companies and individuals establish networking opportunities, looking ahead to leverage the local economy. Bizfeira is available to disseminate the know-how and high quality products for businessmen.

The participants can ask questions on matters related to the town hall directly to the president of *Setúbal* Municipality [19] getting answers in real time. The "chat" is held on Mondays, from 15 to 15 days between 21h00 and 23h00, and it is accessible on Facebook via a link provided on the website of the Municipality. This service is a form of instant conversation through the Internet, text messages, and other means of contact between the President of the Municipality and the citizens.A municipality with interest and the opinion of its citizens encourage the use of new information and technology.

4.3. CIM Cávado and Shop Mobile - Palmela

The *Intermunicipal Community Cávado*[20] is the NUT III *Cávado*, and it includes the municipalities of *Amares, Braga, Barcelos, Braga, Terras de Bouro and Vila Verde*. The CIM *Cávado* aims to combine, promote and articulate common interests of associated municipalities in the area of public proximity services, and municipal investments in terms of: Promoting the planning and management of economic development strategy, social and environmental dimensions of territory covered; Articulation of municipal investment inter-municipal interest; Participation in the management of regional development support programs, particularly under the National Strategic Plan (Quadro de ReferênciaEstratégico Nacional – QREN); Planning of the actions of public authorities, in supra-municipal character.The Municipality of *Palmela*[21] offers the first Citizen Shop Mobile of the country. It consists of a mobile vehicle that circulates through the county, especially in rural areas in order for residents with limited movement can use the services and receive information from the municipality. In the mobile vehicle it is possible to treat the same subjects available in branches of municipal service like: urbanism, water, Advertising and occupation of public space, hygiene and health surveys, food stamps the scope of the feeding program and school transport, information and customer support, licensing public entertainment events, presentation of opinions, suggestions or complaints, Information on conditions for urgent action taking place in the county, information about the cultural offer, sports, tourism and leisure and others.

4.4. Other examples

InTable 1 an overview of the most relevant projects in Portugal and some examples internationally. The examples are from 2005 until the present.

Table 1 - Examples of good practices

County / City	Year	Name	Description
Washington Mercer Island[22]	2006	MyBuilding Permit.com	Project developed to provide services of housing construction. It provides authorization and information for government agencies in the city.
Several states of Washington[23]	2007	NWMaps	NWMaps gives access to virtual information of city maps. It contains information about business and houses.
Washington Sammamish[24]	2005	Sammamish 21 TV	An exclusive TV channel was created for the city. The current schedule shows, the meeting of the town hall, information spots about the city and other programs.
Germany Hamburg[25]	2016	HamburgGa teway	It provides a point of access for all citizens to online services of the city and the Hamburg state. With its strong authentication, ensures privacy and security for the customer.
New Zealand Auckland[26]	2007	Unlocking Auckland's Diversity	A platform has been created to host communities and migrants can work together. This aims to build the city's ability to be welcoming.
Portugal Guimarães[27]	2007	GmrTV	Despite have not been developed by CMG, but the project has their support. It presents the news and events that occurred and will occur in Guimarães.
Portugal SMAS Oeiras and Amadora[28]	2009	Walkthroug h Get New Skills	This project won the award for good practice in local government. The objective was improving the academic knowledge of the workers who work in the various municipal services.
Portugal Seixal[29]	2009	Seixal Digit al - Online Services	The *Seixal* presented the project to put its online services always available. It aims to progressively dematerialize and reduce administrative costs of the services
Portugal Arraiolos[30]		Municipio GIS portal Arraiolos	It was built using open source software. The GIS portal allows the city a wide autonomy in its management of the portal because it has a big flexibility in terms of content.
Portugal Évora[31]	2006	Station Access	The city is equipped with 8 vehicles that travel in the *Évora* district providing to citizens a free access to the internet.
Portugal[32]		Public displays TOMI	These screens are in different cities and they are aimed at providing information to citizens and publicize events. Each municipality has control of the information. The information is based on news, directory (with city information), calendar and transports.
Portugal Trofa[33]	2016	GO – Gestão de Ocorrências	The municipality makes available as of April 1st the platform named "*GO – Gestão de Ocorrências*", to all municipals. This platform allows the warning, signaling, lifting and solving quickly the occurrences in the Trofa municipality. The e-government platform is the first in Portugal available as mobWeDo.Oc in App Store and Google Play Store, which allows to the residents a participatory performance in a network. The network will allow the citizens to cooperate with the municipality in the identification of possible occurrences. For example, the level of building works, abandoned animals, sanitation, waste, abandoned vehicles, and other areas.

5. Discussion

The technological evolution leads to an inevitable growth and development of online applications and services in all society sectors. The governmental sector is no exception. Currently, the services of public and local administration sectors are focused on digital services.

In this paper, it is verifying a possible set of good practices that are applied by municipalities in Portugal. All municipalities have internet access, but the online services still have problems. The good practices already implemented are not always used by the citizens with the expected frequency. Each municipality presents only a reduced set of good practices, because they do not reuse the good practices implemented in other municipalities, which could be used to improve their services. Usually, they only use the good practices and ideas created by them. This work wants to change this paradigm by making a study of the best practices and disseminating it to the community.

6. Conclusion and Future Work

In the last ten years, there was a huge advance on the services implemented in local administration to bring the citizens closer and attract visitor for their municipalities. The proposal of this study is to obtain an overview of eGov in Portuguese local government. For this is necessary to do a deep search about good practices which there are in municipalities and combine them with digital marketing strategies to increase their divulgation.The combination of these two factors will provide a new vision of eGov in Portugal. Furthermore, the study provides to municipalities a guide of good practices which can be implemented by them. These good practices have the proposal of appeal to the citizens to use the ICT and allow a cost reduction by the municipality through new innovative ideas.

The main objectives which lead to adopt digital strategies by the municipalities are the need for documentation dematerialisation, reducing costs and simplifying the demand and supply process. Therefore, a new challenge emerged focused on how the information is treated and how it reaches the citizens. As a result, the Digital marketing methods are being increasingly used due to ICTs evolution which allows the information to be reachedby citizens through the multiplatform. The objective of this work was to identify a set of best practices and disseminates them to society and the scientific community. This article presents the first stage of this goal. The next steps include the analysis of more eGov good practices in Portugal and internationally, defining requisites and restrictions of the services. At the same time an interaction with various municipalities will be scheduled in order to improve their services. A set of indicators also will be defined to be used in a benchmarking process which will be applied in the municipalities which implemented eGov practices. Finally, in this research we developed a guide of good practices which could be analysed and implemented by the Town Halls.

Acknowledges

This work has been supported by Compete: POCI-01-0145-FEDER-007043 and FCT - Fundação para a Ciência e Tecnologia within the Project Scope UID/CEC/00319/2013.

7. References

[1] Community M of 2003 A GUIDE TO MUNICIPAL Meeting the Requirements of the *A Guid. TO Munic. Prog. Report.* 37
[2] Georges K 2005 Uclg country profiles **0** 9
[3] Sharma S K and Gupta J N 2003 Building Blocks of an E-Government *J. Electron. Commer. Organ.***1** 34–48
[4] Zhao F, Collier A and Deng H 2014 A multidimensional and integrative approach to study global digital divide and e-government development *Inf. Technol. People***27** 38–62
[5] Missouri 2015 2015 Missouri Municipal League INNOVATION AWARDS (Missouri)
[6] United Nations 2009 Good pratices and innovations in public governance
[7] Cook M E 2011 Making a Case for Local E-Government 1–16
[8] Teixeira R, Afonso F, Oliveira B, Machado J and Abelha A 2015 Decision Support in E-Government - A Pervasive Business Intelligence Approach *New Contrib. Inf. Syst.*

Technol. **354** 155–66
[9] Teixeira R, Afonso F, Oliveira B, Portela F and Santos M F 2014 Business intelligence to improve the quality of local government services: Case-study in a local government town hall *KMIS 2014 - Proc. Int. Conf. Knowl. Manag. Inf. Shar.* 153–60
[10] Ryan D and Jones C 2009 Understanding Digital Marketing: Marketing Strategies for Engaging the Digital Generation vol 10
[11] Stats W I 2015 World Internet Users Statistics and 2015 World Population Stats
[12] UMIC 2011 Utilização de TIC na Administração Pública Central, na Administração Pública Regional (Açores e Madeira), e nas Câmaras Municipais
[13] European Commission 2014 Delivering the European Advantage? "How European governments can and should benefit from innovative public services"
[14] Brasil D X and Bastos J M 2006 *Manual de Técnicas de Marketing* (Lisboa: CE-PALOP)
[15] Rahman H 2010 Comparative E-Government **25**
[16] Dias G P 2014 Bibliometric Analysis of Portuguese Research in e-government *Procedia Technol.* **16** 279–87
[17] Guimarães C M 2015 Melhor Cidade Digital
[18] Feira S M 2014 Bizfeira
[19] Setúbal C M 2005 Conversas com a presidente
[20] Cávado V 2008 Comunidade Intermunicipal do Cávado
[21] Palmela C M 2012 Loja Móvel do Cidadão
[22] Island M 2006 MyBuildingPermit
[23] Washington 2007 NwMaps
[24] Sammamish 2005 Sammamish 21 TV
[25] Hamburgo M 2016 HamburgGeteway
[26] Auckland M 2007 Unlocking Auckland's Diversity
[27] Guimarães 2007 Guimarães TV
[28] Cimas O A 2009 Passo a passo adquira novas competências
[29] Seixal M 2009 Seixal digital
[30] Arroiolos M 2008 Portal SIG
[31] Évora M 2006 Posto móvel de acesso à internet
[32] Anon TOMI World
[33] Trofa M 2016 MobWeDo

Potential fraudulent behaviors in e-procurement implementation in Indonesia

S N Huda[1], N Setiani[1], R Pulungan[2] and E Winarko[2]

[1] Department of Informatics, Universitas Islam Indonesia, Yogyakarta, Indonesia
[2] Department of Computer Science and Electronics, Universitas Gadjah Mada, Yogyakarta, Indonesia

E-mail: sheila@uii.ac.id, novi.setiani@uii.ac.id, pulungan@ugm.ac.id, ewinarko@ugm.ac.id

Abstract. Corruptions in public procurement have occurred in various parts of the world, especially in developing countries. Implementation of SPSE (electronic procurement system) as the e-procurement system in Indonesia is based on the government's intentions towards clean and good governance by fighting corruption, collusion, and nepotism. Procurement in Indonesia is carried out through SPSE, which is developed by LKPP (Government Policy on Procurement of Goods/Services). Although this system has brought many positive effects, there are still found many practices of fraud occurring in the implementation of the system. In this paper, we try to identify these practices and then to systematically categorize and analyze them.

1. Introduction

Burton [1] argues that public procurement is a core instrument to achieve efficient management of public resources. It supports jobs and government services and can cover a variety of needs, ranging from stationery, furniture, temporary office staffs, to a very complex and high-cost projects, such as construction, aircrafts, and other various projects.

Corruptions in public procurement have occurred in various parts of the world, especially in developing countries. Corruptions have negative effects on various levels: local, regional, and national [2]. Above all, corruptions adversely affect the competence of public and prosperity in a country, increase the cost of government operations and social structures, undermine confidence in the government, and unbalance the composition of government spending on a wide range of services, including education, health, operation, and maintenance. Thus, corruption is a lethal socio-economic problem, and unfortunately, is widespread in developing countries like Indonesia.

Sistem Pengadaan Secara Elektronik (electronic procurement system, abbreviated SPSE in Indonesian) is the decentralized e-procurement system of Indonesia developed by LKPP (Policy Institute for Government Procurement of Goods/Services). SPSE is run by LPSEs (Electronic Procurement Services). Now the number of LPSEs has reached 635, where the process of growing the number of LPSEs is a process that takes time and effort. It mainly depends on the political will of the local governments and on the awareness of the needs to implement e-procurement.

Implementation of SPSE is based on the government's intentions towards clean and good governance by fighting corruption, collusion, and nepotism. Corrupt practices are known to always occur in the practices of government procurements of goods and services [3]. According to a survey conducted by Indonesian Corruption Watch (ICW), which was presented to the Corruption Eradication Commission (KPK) in March 2011, 89% of companies supplying goods and services to the government had bribed to win tender offers [4]. On the other hand, companies making efforts (trying) to offer bribes reached 92%. The survey was conducted on 792 providers of goods and services in Jakarta, Bekasi, Tangerang, Depok, and Bogor [4]. ICW also reported that there are other corrupt practices in the procurement process of goods and services, which include 48 cases of mark-ups, 50 cases of illicit letters, 1 case of breach of contract, and 8 cases of false procurement projects [5].

The presidential decree No. 54 of 2010 states that e-procurement should be applied starting in 2012. The implementation of e-procurement is expected to reduce corruption because all processes of procurement tender are conducted online using the Internet, and thereby reducing the possibility of direct meetings between potential suppliers and procurement organizers.

In general, the implementation of e-procurement in Indonesia brings significant positive results [3]. Among the positive impacts of the implementation of e-procurement in Indonesia is increasing the efficiency of fund procurement, and increasing procurement transparency to reduce corrupt practices in order to achieve clean and good governance. However, few researches investigate frauds in the implementation of e-procurement itself. Although it has brought many positive effects, there are still found practices of fraud occurring in the implementation of e-procurement. For example, as disclosed by the head of LKPP, Agus Rahardjo, in 2014, LPSEs of Malang, Banten and Palembang have been given warning letters from LKPP citing deliberate intentions to block some companies or providers from getting into the tender process of procurement programs, so the winner of the tender process is some preferred or targeted partner company (http://www.lkpp.go.id/v3/#/read/2869, 25 September 2014).

Nurmandi [6] suggests that the status of e-procurement in Indonesia is in the middle of the status quo where until 2012 the number of procurements auctioned electronically by the central government only reached 10.26% of the total procurements and 21.10% of the total procurements are by local governments. Based on LPSE smart report in 2015 for 6 years of SPSE implementation, the total number of LPSEs increases from 10 with 33 total tender offers in the initial year to 570 LPSEs with 135.669 total tender offers and 129.728 finished tenders. These LPSEs rendered the tender processes more efficient by more than 17.9 billion Rupiahs, which is equivalent to 8.41% of the total budget.

Although the implementation of e-procurement has improved the transparency of the procurement process, some people are still able to exploit limitations of the system, and this in turn still allows for corrupt practices to occur. This paper discusses potential fraudulent behaviors that have occurred in the implementation of e-procurement in Indonesia.

2. Methodology

The method used in this paper is a qualitative research method with the primary method is to obtain data through interviews. Parties who were interviewed in the study include LKPP, several LPSEs, and providers of goods and services that have followed the tender process via e-procurement. Data collection was performed using purposive sampling, *i.e.*, by selecting individuals who have the expected data. We interviewed LKPP, provincial, district, and state higher education institution LPSEs, as well as providers of goods and services. After obtaining qualitative data, we analyze data using open coding thematic analysis method. Interview results will be identified as themes of occurred fraud cases that potentially threaten the implementation of e-procurement.

3. Potential fraudulent behaviors

Frauds in the e-procurement application can be categorized into behaviors that involve people from inside and outside of an LPSE. These behaviors are summarized in table 1.

Table 1. Insider Frauds and Outsider Attacks.

Insider

No.	Fraud	Common Cause
1	Bandwidth limitation	No systematic monitoring;
2	Firewall rule intervention	Diverse security levels of
3	File changes/removals by system admins	decentralized LPSE servers;
4	Subjectivity on assessment	Lack of personnel's integrity.

Outsider

No.	Attack	Common Cause
1	Hacking	Lack of security awareness;
2	Instrusion	Diverse security levels of
3	Potentially dissatisfied ex-system admins	decentralized LPSE servers;
		Lack of personnel's integrity.

3.1. Fraudulent behaviors by insiders

LKPP has issued standardizations for LPSEs whose components include availability, security and service. Security aspects are adopted from ISO 27001 with adjustments. These includes server placement in integrated data center, data center physical security, such as CCTV and finger print usage, recorded and limited access, and SSH regulation. However, this standardization is not mandatory for all LPSEs, resulting in a wide variation of LPSE server condition. Because of this variation of LPSE server condition, server security awareness also varies. Based on our research, several methods of frauds involve the procurement committees, including:

3.1.1. LPSE network/server bandwidth limitation This is usually done at the time of submission. At the time when the preferred provider has successfully input the bid file and all necessary documents have been uploaded 100%, server's bandwidth is reduced so that other potential providers are not able to upload their documents. Several LPSEs have been caught and given reprimands for committing a network cheat. The lure of cheating providers may attract system administrators with weak integrity.

On the other hand, government has to consider the welfare of the committees, especially system administrators, who hold great responsibility in managing e-procurement system. The welfare of system administrators, in the form of performance benefits, varies greatly among regions, depending on the LPSE position, along with the magnitude of the local budget. In LPSEs under Diskominfo (Government Office of Communications and Information Technology), their remuneration has been considered properly. But ad hoc LPSEs usually provide remuneration that is not as good as other areas. This can lead to a potential lack of integrity on the part of committees/system administrators, as they have similar workload between one region and others, but receive different remuneration. System administrators' quality is based on the aspects of skill, attitude, loyalty, and morale. In LPSEs located in a big city, it is easy to find system administrators with high skills. In remote areas, however, it is more difficult to find system administrators with good skills, let alone to filter their attitude, loyalty, and morale.

3.1.2. The use of firewalls to inhibit the tender process This is performed, for example, by creating a rule in an LPSE firewall, where key to the uploading document page is locked so that some providers' uploading document page is empty and potential providers therefore cannot upload their bidding documents. Network and firewalls are vulnerable because servers are managed by decentralized LPSEs.

In some LPSEs, distribution of roles exists to avoid network abuse. In East Java province's LPSE, for example, this role is divided into three institutions: the management of LPSE server in data center under the Office of Communications and Information Technology (Diskominfo) of East Java province, administration system under the Bureau of Development Administration, and UPT P2BJ as the tender committee. Hence, server management is not directly under the direct procurement committee. This restricts the tender committee from cheating on the network and firewall. Data center in Diskominfo of East Java province has met safety standards and the office expects each district/city in East Java province to put their LPSE servers in the provincial data center. However, there are many obstacles that make districts/cities reluctant to entrust their servers to the provincial data center. The sectoral ego between different government agencies is still high. Even with certified security standard ISO 27001 data center, some institutions are not willing to entrust their LPSE servers to the provincial data center.

For other cases, South Sumatra province's LPSE has been reprimanded for indications of frauds in the implementation of e-procurement several years ago. It became very difficult to restore public trust, so the LPSE chose to entrust its server to a competent and professional certified third party. This third party was also under direct supervision from LKPP, reducing potential misuse of the server and the network. However, this will relate to the dignity of government institutions, which in this case is Diskominfo as an institution that acts as the government's Chief Information Officer (CIO). This institution will be deemed incompetent to manage data center or LPSE server.

One solution to overcome the reluctance caused by sectoral ego and to restore the dignity of Diskominfo as CIO of government is to build data centers managed by Diskominfo, but to place them on a higher institution. For the province level, for instance, the server can be put in the governor's office, or for the district level, it can be put in the regents office, or other close location as long as it holds certified and standardized security. This requires a willingness from local leaders in order to minimize sectoral ego. On the other hand, LKPP actually has developed a program to record the activities of LPSE servers. But it is only limited to recording, and yet is not able to analyze the activity. The number of records obtained from 635 LPSEs is certainly very large, making analysis very difficult. So far, the record log is only used if there is a complaint so that research in this area is certainly required.

3.1.3. File changes by server administrators Changing files as well as the destruction or removal of files, such as bidding files, are intended to omit other potential providers' bidding documents, so that those potential providers will be void. The perpetrator will usually also erase the server access log. Moreover, some LPSEs have improper server rooms, in the sense that many people can access the server freely. Based on our research, some SKPDs (working units) put LPSE on a server that can be accessed by strangers, not physically safe, on the floor, or on a pile of boxes that are susceptible to physical damage. Even one LPSE put the server in the bidding room, where bidding room is accessible to just anyone, including potential providers who wish to upload bidding files locally. Not only physical damages, the risk of theft is also prominent.

3.1.4. Subjectivity in the assessment of technical and qualification evaluation During the evaluation of the bid proposal, price, qualification, and technical assessment are evaluated. For the evaluation of the price, since it is nominal, it can be ascertained objectively. However, for the qualification assessment and technical assessment, there is still a factor of subjectivity. For a single potential provider, one appraiser evaluation committee may assume that the provider has similar experience with the work being tendered, but it is also possibly judged as dissimilar by another committee; hence results in a minimum rate. Definition of similar or dissimilar yet is unclear. As another example, suppose there is a tender for an information system. It is currently possible that the appraiser of technical proposal has no background in information technology. Different backgrounds, such as economics and informatics, will have different way in assessing the technical proposal, resulting in valuations that are less objective.

3.2. Fraudulent behaviors by outsiders

3.2.1. Hacking by irresponsible people Some mischiefs from hackers, such as DDOS, malware, and backdoor, have occurred. Firewall setting is very important to prevent DDOS attacks. In some LPSEs DDOS attack did occur. The DDOS attack could be generally attacking the whole system in the domain (for example, attacking one district), or considered deliberately attacking only SPSE to impede the tender process, especially to make the system down and potential providers then cannot upload their bidding documents. In terms of data communication, since the SPSE application is a web-based application, it should use HTTPS (SSL) by default. From our observations, only about 20-30% of all existing LPSEs have taken advantage of the HTTPS protocol on their SPSE system. The rest, 70-80% of them, still use the HTTP protocol only [7].

3.2.2. Intrusion or infiltration Intrusion is performed by persons who are not permitted to gain administrator privileges. As reported in http://www.antaranews.com/berita/554513/bareskrim-usut-kasus-pembobolan-akses-lpse-kemen-pupr, and some other news portals, the intrusion problem had raised security problems in LPSEs. On infiltration, when permissions as system administrator is obtained, the perpetrator can change individual files, removing files, change the server's time, and so on. Disappearance of files has recently occurred in several LPSEs, which is caused by hackers who deliberately remove bidding files and leave only one file belonging to a partner provider who paid the hackers. It has been categorized as a criminal conduct and has been under criminal investigation.

3.2.3. Former system administrators System administrators who have resigned, or moved to other institutions, but still have access to the SPSE servers have a great opportunity to commit crimes. As an administrator who has access to the server, he can change or remove files. In various government agencies, there is no clear standard operating procedure for the mutation or resignation process of system administrators.

3.2.4. Information leakage Information leakage is basically not categorized as attack or fraud, but has potential to weaken SPSE. There is a very complete document standard operating procedure (SOP) of LPSEs published online. Some information from the SOP is information that should not be shared publicly, such as:

- SPSE installation process and detailed SPSE framework.
- Web configuration, path configuration, database configuration, and system/application log file.
- List of application packages and supporting applications, application paths and application versioning.
- Web security configuration, web application firewall usage, and other security modules.
- Database name, user and password used in training and production phase.
- Monitoring log instructions, data center safety, backup and recovery information.
- Security, infrastructure, and system incident handling procedures.

This information can be misused by hackers by making it easier for them to penetrate the SPSE security.

4. Conclusion and future work

Potential fraudulent behaviors in the implementation of SPSE can be categorized into two groups: frauds involving insiders and frauds in the form of attacks from outside. Computer system security can be divided into several layers: physical security layer, human resources (people) layer, data/communications engineering (system/technology) layer, and policy & procedure layer [8]. To handle human resources layer, the government needs to pay attention to the system or network administrators, to standardize their skills and remuneration. To handle physical security layer, government also needs to pay attention to the physical security of servers where the LPSE servers still have varied qualifications. The use of datacenter with high security standard needs to be done, at least in one province. Regencies/cities that have no adequate datacenter can entrust their LPSE servers to the province data center. Strengthening the security in technology layer, such as by cloud LPSEs, can be studied further. The last dimension, *i.e.*, the policy requires a study of information system management system (ISMS) implementation, which has been initiated by the Ministry of Communication and Information Technology, as soon as possible, considering that e-procurement is a strategic e-government service.

Acknowledgments

The authors would like to thank the Ministry of Research, Technology and Higher Education of Indonesia for funding this research.

References

[1] Burton R A 2005 Improving integrity in public procurement: the role of transparency and accountability *Fighting Corruption and Promoting Integrity in Public Procurement* (OECD Publishing) pp 23–8
[2] Ampratwum E F 2008 The fight against corruption and its implications for development in developing and transition economies *Journal of Money Laundering Control* **11** 76–87
[3] Setyadiharja R, Budiman S, Karim Z, Matridi R, Junriana J, Ferizone F and Nurmandi A 2014 E-procurement system technology: an analysis in electronic procurement service unit (LPSE) of Kepulauan Riau Province *The Asian Journal of Technology Management (AJTM)* **7** 93–107
[4] Kredibel 2011 E-procurement innovation towards free corruption procurement *Kredibel Procurement Magazine* **Oct.-Dec.** 12–3
[5] Purwanto A, Ibty I, Rofikah N and Indroyono P 2008 E-procurement in Indonesia (development services procurement electronic government) *Kemitraan Partnership dan LPSE Nasional*
[6] Nurmandi A 2013 What is the status of Indonesian e-procurement? *Journal of Government and Politics* **4** 350–73
[7] Sinambela J 2016 LPSE [in]security *Indonesia's Information Security Portal* URL `http://infosec.id/2016/05/lpse-insecurity/`
[8] Icove D, Seger K and VonStorch W R 1995 *Computer Crime: A Crimefighter's Handbook* (O'Reilly & Associates, Inc.)

Power Exercise in the Institutionalization of an eProcurement System in Indonesian Local Government

Fathul Wahid

Department of Informatics, Universitas Islam Indonesia, Yogyakarta, Indonesia
fathul.wahid@uii.ac.id

Abstract. Power possessed by institutional entrepreneur is believed to have an influential role in navigating the institutionalization process of an idea. However, only few studies that have examined how the institutional entrepreneur exercise his power throughout the institutionalization process. This study aims at examining this phenomenon in the context of developing countries with special reference to the implementation of an eProcurement system in Indonesia. It finds that the power has been exercised by the institutional entrepreneur differently along the process. Smart power was exercised by combining soft power at the early stage of institutionalization process. At this stage level of reluctance among organizational members was high, no authorization from higher-level agencies, and the benefits of adopting the system were opaque. Hard power was then exercised at the later stage, when the legal support was well formulated, and the system's benefits were confirmed. The study also reveals that value-based power that is used wisely make organization change runs smoothly until it becomes institutionalized. Keywords: Institutionalization, smart power, soft power, hard power, value-guided power, eGovernment, eProcurement, Indonesia, developing countries.

1. Introduction

Literature documented two streams of research on organizational changes. The first stream stresses naturalistic evolution, whereas the second, agent-based view, emphasizes power and intentional design [1]. Moreover, previous studies acknowledge that the role of institutional entrepreneurs in initiating and leading divergent organizational change that makes an idea becomes implemented and institutionalized [2-4]. Other studies point out that their power is important in this institutionalization process [1, 3, 4].

Power is often associated with domination, and in many cases is responded with resistance [5]. Resistance from organizational members has been frequently cited as one of main obstacles in organizational change. Hence, identifying strategies on how to change an organization smoothly without drama is then important [6]. However, only few studies that go further by examining how the power is exercised during the institutionalization process [7, 8]. There is also a call that researchers need to power back in institutional theory [1, 9].

This study aims at examining how the power of the institutional entrepreneur has been exercised during the institutionalization process of an eGovernment initiative. Without neglecting the role of other actors, in this study, the institutional entrepreneur in question is the dominant one, i.e. the mayor of the city. It attempts to address this main question: *how does the institutional entrepreneur exercise his power throughout the institutionalization process?*

To answer this question, an interpretive case study was conducted to trace back the institutional process of an eProcurement system in an Indonesian city of Yogyakarta. Generally, institutional theory is used as lens, and specifically this study focuses on the concepts of institutional entrepreneurship and power [1]. This study is important as it explicates an example of successful eProcurement projects in the context of developing countries, which experience a low success rate in implementing eGovernment initiatives in general [10] and eProcurement programs in particular [11].

The rest of the paper is organized as follows. After this introduction, Section 2 presents the conceptual premises of this study. Section 3 describes the context and research methods, followed presentation of the findings and discussion in Section 4. Section 5 concluded this paper by summarizing the findings and identifying limitations and possible future research.

2. Theoretical Basis: Institutionalization and Power

There are two research streams in examining institutionalization. The first stream focuses on institutionalization as an outcome (i.e. an institution), while the second focuses on the process of institutionalization [12]. As noted earlier, the second has received less attention in the literature [7, 12]. Thus, this study focuses on the latter.

Institutionalization can be seen as a process of instilling values [13]. Before a value is instilled in an organization, it only has instrumental utility as a mechanical and disposable tool [14]. The institutionalization process is often preceded or accompanied by a deinstitutionalization process, which is "the process by which the legitimacy of an established or institutionalized organizational practice erodes or discontinues" [15:564]. The institutionalization process may be triggered by environmental jolts (i.e. changes at the organizational field) or institutional entrepreneurs who recognize problems and lead the deinstitutionalization process [3, 7, 16]. The old institution is disrupted and the new institution is created and maintained [17].

Institutional entrepreneurs have been defined as "change agents who initiate divergent changes, that is, changes that break the institutional status quo in a field of activity and thereby possibly contribute to transforming existing institutions or creating new ones" [2:76]. Here, an actor can be an organization or a group of organizations, an individual or a group of individuals [2]. This study pays special attention to a dominant individual, i.e. the mayor of the city. Institutional entrepreneurs are purposeful actors who have the necessary, resources, social and political skills at their disposal. In addition, they are in a social position (in this study the position as the mayor) to be able to lead change [2, 18]. Institutional entrepreneurs are actors who initiate divergent changes (i.e. changes that break with the existing dominant institutional logic or the institutional status quo), regardless of whether or not they succeed in their initiatives [2].

Institutional entrepreneurs act on a variety of motivations: for example, these may relate to material wealth and status, values, ideology, power or problem solving [2]. They serve as actors of legitimacy, supporting the creation of institutions that they consider to be appropriate and aligned with interests they value highly [18, 19]. These interests will influence institutional logic, and the character of the new institutions will be dependent on them [2]. Resources are integral to institutionalization; thus, institutional entrepreneurs use strategies to mobilize these resources [18, 19], which cannot be separated from the use of their power. Power is a relational concept: one can have it only in relation to others [8]. For the purpose of this study, power is seen as resulting from access to and control over resources that consist of cognitive, social, and material ones [18].

The extant literature defines and puts meaning on the concept of power differently [8, 9, 20, 21]. Issues of power and influence were central in the old institutionalism [22], but they received less attention in the new one [1]. Power has been both an independent variable causing outcomes such as influence, domination, and accumulation of valued resources; and a dependent variable, the results of, such as, dependency and centrality [8, 20]. In a review of 82 papers concerning the relationship between power and information systems, Jasperson et al. [21] identified four lenses for examining power: (a) rational – focuses on authority, information, and expertise as its basis; emphasizes rational decision making; and focuses on single goals, (b) pluralist – does not assume a single goal; assumes that the organizational members have different, potentially conflicting interest, (c) interpretive – focuses on symbolism, metaphors, language and meaning; deals primarily with perceptions and the processes that shape them, (d) radical – focuses on discussion of power and information systems in a larger social context.

Power can be differentiated based on the way it is used in a context: (a) hard power is used by coercing others, (b) soft power is exercised by persuading others, and (c) smart power is seen "as the capacity of an actor to combine elements of hard power and soft power in ways that are mutually reinforcing such that the actor's purposes are advanced effectively and efficiently" [23:115]. Smart power requires recognition a variety of institutional forms and relationship across among the organizational subunits/members, all of which have their own values and rigidities [23]. Often, power exercise is seen as a way to solve resistance during the process of organizational changes [5]. Inline with the pluralist view of power, Fleming and Spicer [5] suggest to go beyond this diametric relationship, by introducing the concept of struggle. They argue that this concept could provide a term for thinking about power and resistance as an interconnected dynamic.

At least three distinct resource mobilization strategies can be identified: convening, leveraging and accumulating. These three strategies are not mutually exclusive and may be conducted simultaneously [24]. In the convening process, the mobilization of resources is carried out through collaborative arrangement as a precondition for institutional change [24]. This involves 'collective institutional entrepreneurship', namely "the process of overcoming collective inaction and achieving sustainable collaboration among numerous and dispersed actors to create new institutions or transform existing ones" [25:1079]. This study gave as an example of the convening process, a situation in which the mayor attempted to gain support from other actors in the initial phase of the implementation of an eProcurement system. At that time, he used a normative approach rather than coercive power. In the leveraging process, politically skilled institutional entrepreneurs act as the driving force [24]. They mobilize legitimacy and support the use of frames to shape people's perceptions, cognitions and preferences [2]. An example from this study is a situation in which the mayor, supported by his backers, exercised his power to gain more support and resources to further implement an eProcurement system. The mayor took this course of action after the benefits from adopting an eProcurement system had become apparent. In the accumulating process, support is gained through the actions and interactions of various actors. These result in a dominant design, which is then diffused [24]. This situation happens, for instance, when there are multiple and conflicting institutional logics or when more than one key institutional entrepreneur is involved. Their actions and interactions, over time, result in dominant patterns, which then are widely accepted.

Several indicators of institutionalized practices exist: they are generally accepted without debate by the organizational members, are widely followed, exhibit permanence [26], and their existence is no longer dependent on powerful actors [27]. Hence, institutionalized practices become valued organizational routines and an integral part of day-to-day practices. According to Jarzabkowski [28], practices can be seen as "repetitive performance in order to become 'practised'; that is, to attain recurrent, habitual, or routinized accomplishment of particular actions".

3. Research Setting and Method

3.1. Setting the Stage

eProcurement was formally adopted by the Indonesian government in April 2007. One of the strategic objectives of this initiative is to improve the transparency and accountability of public procurement. This objective has gained ground in Indonesia because budget leakage in public procurement has reached an alarming level. Several sources indicate that between 10% and 50% of the public procurement budget has been misappropriated [e.g., 29, 30]. This is a huge number since public expenditures account for 30% to 40% of total national spending in Indonesia [31].

Corruption that affects public procurement in Indonesia involves a wide spectrum of individuals and organizations, including political leaders, judicial figures, senior administrators and officials in procurement roles, as well as suppliers and contractors [32]. Therefore, it is important to have sound processes and proper management to reduce and minimize the potential for irregularities and misconduct [31].

The research site, namely the city of Yogyakarta, implemented eProcurement in 2008. However, preparatory activities began in 2007. By the end of 2008, 11 work packages had successfully been completed with the help of an eProcurement system. The work packages tendered were valued between IDR 500 million and 1 billion to attract a large number of vendors. In 2009, all work packages with a value greater than IDR 500 million (51 work packages) were tendered via the system. At the beginning of 2009, Yogyakarta was described by Transparency International Indonesia as the cleanest city in the country in terms of its anti-corruption practices. Since 2010, the eProcurement system has been used for all tenders valued at more than IDR 100 million. Up until the end of 2015, 1,858 tenders were made using the eProcurement systems, offering a saving of 13.66% (around IDR 295 billion) (see Table 1).

Table 1. The use of the eProcurement system.

	2008	2009	2010	2011	2012	2013	2014	2015
Number of tenders	11	51	137	247	342	300	432	342
Value (IDR million)	5,449	78,866	141,267	249,278	376,918	367,696	391,517	338,268
Efficiency (IDR million)	225	12,649	22,404	37,567	52,824	52,824	52,824	52,824
Efficiency (%)	4.34	16.18	16.10	15.07	12.33	17.19	10.86	12.02

Source: Interviews and http://report-lpse.lkpp.go.id

3.2. Research Method

This case study is interpretative in nature. The data were collected data from various sources: interviews, documents, and observations.

Interviews were conducted with variety of key players in eProcurement or eGovernment implementation at various levels. The interviewees were the mayor of the city of Yogyakarta, four heads/vice-heads of departments, three heads of divisions and three administrators. The snowball method was used to select the interviewees. Some of them were no longer involved, such as the former head of the IT division who was later assigned to different tasks not related to the eProcurement initiative. He was the first informant who pointed to other relevant informants. In addition, the existing organizational structure was consulted to identify potential informants. Involving key players at different levels as research informants was intended to make better sense of the phenomenon under investigation [33] and to avoid elite bias [34], i.e. overweighting data from articulate, well-informed, usually high-status informants and, conversely, under-representing data from less articulate and lower-status ones. In total, 11 semi-structured interviews were conducted, which were between 30 and 150 minutes long. Most of the interviews were recorded. Otherwise, note taking was conducted.

In order to ensure the validity of the interpretation of the phenomenon, whenever possible, member checks [33, 35] were conducted, mostly through an informal discussion with three of the interviewees. In these discussions, additional information for some areas, which were not obvious during the previous interviews, was obtained.

In addition to the interviews, data were also collected from written documents/reports, websites, presentation slides, and news presented in the media. Field observation, including participating in a national meeting of eProcurement agencies, was also made. The intention was to better understand the context of the study by collecting information about the implementation of eProcurement system in other districts and at the national level agencies. The understanding was very important when sensemaking the data.

In analysing the data, various concepts of power (especially soft and hard power) were used as templates in the coding process. Since the notion of power holds multiple meanings, here, when referring to it, this study included authority, decision, policy, and influence. In this study, as top management, the mayor certainly has formal authority to collaborate with other government agencies (such as with the National Development Planning Agency (*Badan Perencanaan Pembangunan Nasional* [Bappenas] and the Corruption Eradication Commission (*Komisi Pemberantasan Korupsi* [KPK]), to make decisions (such as intention to adopt the eProcurement system in 2007 and allocating government budget to support it), and to set up policies (formalised in several mayoral decrees) (see Table 2). He has also influence to persuade or attract other actors to become his backers, especially in the early stage of the implementation of the eProcurement system.

Table 2. The critical events throughout the institutionalization process.

Year	Event or status
2007	• The mayor declared his intention to adopt the eProcurement system. • An internal preparation team was setup to learn how eProcurement works by visiting two other cities and the National Development Planning Agency (Bappenas). • The mayor mobilized resources and support from various stakeholders. • In collaboration with the Corruption Eradication Commission (KPK), all the government officers (around 9,000) signed an Integrity Pact (*Pakta Integritas*) to reject corrupt practices.
2008	• Resources were allocated and granted. Support was gathered. • The city signed a memorandum of understanding with Bappenas (6 March). • Mayor's decree on the establishment of an eProcurement Agency (Layanan Pengadaan Secara Elektronik [LPSE]) was released. • The eProcurement system and agency were launched officially (25 July).
2009	• Mayor's decision came into force. All tenders valued at more than IDR 500 million had to use the eProcurement system. • Mayor's decree on the establishment of the Procurement Services Unit (Unit Layanan Pengadaan [ULP]) was released. • The city was described by Transparency International Indonesia as the cleanest city in Indonesia with regard to its anti-corruption practices.
2010	• Mayor's decision came into force. All tenders valued at more than IDR 100 million had to use the eProcurement system. • Mayor's decree on the technical guidelines of the eProcurement mechanism was released.
2011	• All tenders valued at more than IDR 100 million had to use the eProcurement system. • The eProcurement Agency of Yogyakarta was named the Motivating Agency (LPSE Motivator) by the National Public Procurement Agency.
2012	• All tenders valued at more than IDR 100 million had to use the eProcurement system.
2014	• Mayor's new decree on the technical guideline for public procurement to address national government regulation was released.
2015	• Mayor's new decree on the technical guideline for public procurement to address national government regulation was released. • All tenders valued at more than IDR 200 million had to use the eProcurement system.

4. Findings and Discussion

The findings reveals important insights on how the institutional entrepreneurs exercise the power in an emerging field where interactions between actors are not yet stabilized and routinized and they still have a weak mutual awareness [cf. 36]. The eProcurement practices in the context of this study could be considered as an emerging field that successfully challenged the highly institutionalized old procurement system. Table 2 summarizes the critical events throughout the implementation process of the eProcurement system in Yogyakarta, especially when the power is exercised. Other aspects have been reported elsewhere [4, 37].

The findings explicate that the role the institutional entrepreneur, i.e. the mayor, was very influential, as stated by one administrator: *"[The resistance] lasted only for a while. It was dependent on the mayor. At that time [in 2009], the mayor decided that every work package valued at more than IDR 500 million should use eProcurement. Period."*

A head of department also asserted: *"If the mayor is committed, then the process can be accelerated. ... Based on information we got from other districts, many of them are willing to accelerate the adoption of the eProcurement, but their mayor or head of district gives no responses. There are also some opposite cases."*

Despite this fact, this study unearths that the institutional entrepreneur use his power carefully throughout the institutionalization process. At the beginning (2007-2008), the institutional entrepreneur can be said to have a pluralist view of power, as there were multiple institutional logics in play. At that time, some heads of department were still reluctant to use the systems by 'hiding' under the 'collaboration' logic between the procurement officers and the vendors. In practice, this 'collaboration' logic can be considered as the root of corrupt practices. In addition, the mayor and his backers struggled to convince the local parliament, for example by inviting KPK. Hence, at the beginning of the eProcurement system implementation, a convening strategy in resources mobilization was chosen. This strategy was executed through collaborative arrangement [24] and involves 'collective institutional entrepreneurship', that is, "the process of overcoming collective inaction and achieving sustainable collaboration among numerous and dispersed actors to create new institutions or transform existing ones" [25]. This strategy could be regarded as soft power exercise.

Only after the claimed benefits of the eProcurement system were prevalent (in 2009), a more coercive (hard) power was exercised through the used leveraging strategy of resource mobilization. A set of local regulation enacted at the same time also provided deterring effect to those who were reluctant to support and use the system. In connection to this, metaphorically, the dominant institutional entrepreneur (i.e. the mayor) asserts: *"When climbing a mountain, you find a steep track. For me, resistance is not a burden to stop me or to set me back. We only need to know when we should move slowly, when we should move fast. It's more about how to choose the right rhythm to play."*

Above quotation indicates that the power should not be used blatantly, but rather it is exercised carefully by taking the specificity of the context into account. As Fligstein [38] argues, the use of particular strategies by institutional entrepreneurs "depends very much on whether or not an organizational field is forming, stable or in crisis". The mayor's standpoint precisely illustrates the concept of struggle [cf. 5], when power meets resistance.

Here, the institutional entrepreneur is a powerful actor but he uses the power wisely by taking the contextual elements into consideration. This approach is more prevalent at the beginning of the institutionalization process where the institutional entrepreneur adopted normative mode of influence, that is called by Scott [39] as soft power.

The findings do not only point out that power is needed by the institutional entrepreneur, but it goes beyond by providing empirical evidence how the power is exercised. The mayor argues: *"Transform power into 'political waqf'*[1]*. It can be understood in a broad sense: not arrogant, open minded, risk taking. There is always a group of people who support and oppose your decisions. ... In 'the political waqf', we give more than we take."*

Hence, power is only a necessary condition, not sufficient one, for an actor to become institutional entrepreneur. As I observed in some local governments, many powerful actors cannot bring about divergent organizational change, which in this study, drive the institutionalisation process of the eProcurement system. This study reveals that values to which the institutional entrepreneur believes are important in stimulating and driving the institutionalization process. This finding corroborates Selznick's [13] conception of institutionalization which is about value instilling. The mayor puts it this way: *"As I understood it, the eProcurement agency advanced by themselves, with their own innovation without my input. At first, I only cultivated values and shaped them. The hardest part [in the*

[1] The mayor uses the term 'waqf' metaphorically. In the Islamic tradition, the original meaning of 'waqf' is an endowment made by a Muslim to a religious, educational, or charitable cause.

organizational change] is if there are no values. ... An institution is like a box and its foundation is values".

Instilling values is even more important in an emerging field like the implementation of the eProcurement system, since it has to deal with the highly institutionalized corrupt practices in public procurement. Although curbing corrupt practices that have been taken for granted is not easy, but they are not persistence [40]. Instilling values is a part of the efforts.

This study goes beyond mentioning that instilling values is important, but it also reports how the dominant institutional actor (i.e. the mayor) spreads it to other actors, as shared values are regarded as the primary basis of a stable social order [36]. Showing values to other actors is not enough; the institutional entrepreneur should act as role model. This is very important in a context with a culture of great power distance, like Indonesia, where inequality in power distribution is accepted and, in some cases, expected by less powerful people; leaders play a very influential role in deciding 'to go' or 'not to go' [41]. In such a context, considering the institutional entrepreneur, as a 'hyper-muscular actor' is not inappropriate, and this is against criticism on this concept in articulated literature [1, 42, 43].

What can the institutional entrepreneur do in this regard? Leading by example is necessary. The mayor provides an illustration: *"I gave them an example how to use and how I took advantage of technology. They would follow. ... When we were talking about eGovernment, we did not see to the social status. You can send an SMS to a mayor without any burdens. ... I was talking about to have a more effective, efficient, transparent, and accountable work; which was inherent in good governance practices."*

If so, one may ask, why the mayor did not adopt the eProcurement system earlier? Contextual conditions such internal readiness and the absence of supporting regulation, may explain this. Organizational change is a response to the will of institutional entrepreneur; it requires resource mobilization; and it is dependent on the availability of opportunities [24].

Another important issue is identifying under which circumstances either soft or hard power is preferable. The finding presented above provides several insights, which are summarized in Table 3.

Table 3. Soft versus hard power.

Circumstance	Soft power	Hard power
Level of reluctance of the adopters	High	Low
Number of backers	Limited	Many
Authorization from higher level agencies	Absent or weak	Strong
Legal supports	Not ready or opaque	Well formulated or transparent
Benefits of adopting the system	Expected	Confirmed

Taking all into consideration, it seems that the institutional entrepreneur has applied the concept of smart power [23], by combining soft and hard power at different institutionalization stages appropriately. In doing so, the role of values is very important. In this context of this study, the use of smart power could be best described as value-guided power exercise.

By considering the readiness of the context, the mayor exercised soft power in the early stage of implementation of the eProcurement system. At this time, the number of his backers was limited and the resistance was still high due to lack of empirical proof of the effectiveness of the system. Moreover, there was no obligation to adopt the system set by the national government. The legal support was not ready. Instead of waiting for proper legal supports from the national government, the mayor decided to release several mayoral decrees. After the effectiveness of the system was confirmed, the mayor started to use hard power. At this time, there was a strong support from the national government.

This kind of power exercise was proven to be useful in institutionalization of the eProcurement systems smoothly, by minimizing resistance. At the same time, it also increases the legitimacy, as it attracts as many as possible stakeholders to support the system. After all, institutionalization is about making a system widely accepted and taken for granted.

5. Conclusion

This paper has reported how the institutional entrepreneur has exercised power during the institutionalization process. By carefully paying attention to the contextual issues, they opted to use smart power by combing combining soft (normative) and hard (coercive) power. The study reveals several circumstances under which either type power is more appropriate to be exercised (Table 3). In short, in the context of this study, 'Kung Fu masters' are an appropriate metaphor for the institutional entrepreneurs since they use the power smartly and aesthetically, as an opposite to 'Wrestling fighters' who often use the power roughly, though according to a code of rules.

This study has some limitations. First, it paid specific attention to the dominant institutional entrepreneur. Analysing the roles played by other actors, such as heads of departments, will provide a more complete picture of the interplay of power among them. Second, this study inherited all the limitations associated with a single case study, including a lack of generalizability in the sense of statistical sample-to-population inference. However, because it is an interpretive case study, the generalization was attained by extrapolating the empirical findings to theory, either using a supporting theory or a rival one [35, 44]. As such, this study can contribute to theory refinement and development. Third, as a single case study, this study could not capture all possible alternative trajectories of eProcurement institutionalization. In many cases, eGovernment system implementation is context-sensitive [45]. Thus, this study cannot provide alternate explanations for the use of power in a failed institutionalization process, because the system studied was a successful one.

This study offers several contributions. Firstly, it fills the void by responding the call for research on how the power is exercised during the institutionalization process [cf. 7, 8]. Secondly, it offers strategies on how to change an organization smoothly without drama [cf. 6], by identifying a set of circumstances under which smart power is a wise choice. In this context, this study adds the notion of value-guided power into the concept of smart power. Thirdly, it offers insights on how the power is used in the institutionalization process from a less-articulated context in the extant literature, which is the context where highly institutionalized corrupt practice is prevalent from a culturally rich developing country.

References

[1] Clegg S 2010 The state, power, and agency: Missing in action in institutional theory? *Journal of Management Inquiry* **19** 4-13.
[2] Battilana J, Leca B, Boxenbaum E 2009 How actors change institutions: Towards a theory of institutional entrepreneurship *The Academy of Management Annals* **3** 65-107.
[3] Greenwood R, Suddaby R 2006 Institutional entrepreneurship in mature fields: The big five accounting firms *Academy of Management Journal* **49** 27-48.
[4] Wahid F, Sein MK 2013 Institutional entrepreneurs: The driving force in institutionalization of public systems in developing countries *Transforming Government: People, Process and Policy* **7**.
[5] Fleming P, Spicer A 2008 Beyond power and resistance: New approaches to organizational politics *Management Communication Quarterly* **21** 301-9.
[6] Golden-Biddle K 2012 How to change an organization without blowing it up *MIT Sloan Management Review* **54** 35-41.
[7] Mignerat M, Rivard S 2009 Positioning the institutional perspective in information systems research *Journal of Information Technology* **24** 369-91.
[8] Lucas JW, Baxter AR 2012 Power, influence, and diversity in organizations *The Annals of the American Academy of Political and Social Science* **639** 49-70.
[9] Silva L, Backhouse J 2003 The circuits-of-power framework for studying power in institutionalization of information systems *Journal of the Association for Information*

Systems 4 294-336.

[10] Heeks R 2002 Information systems and developing countries: Failure, success and local improvisations *The Information Society* **18** 101-12.

[11] World Bank 2010 Piloting Program in the Use of Country Procurement Systems: Second Progress Report (Washington: World Bank).

[12] Dillard JF, Rigsby JT, Goodman C 2004 The making and remaking of organization context: Duality and the institutionalization process *Accounting, Auditing & Accountability Journal* **17** 506-42.

[13] Selznick P 1957 *Leadership in Administration* (New York: Harper and Row).

[14] Scott WR 1987 The adolescence of Institutional Theory *Administrative Science Quarterly* **32** 493-511.

[15] Oliver C 1992 The antecedents of deinstitutionalization *Organization Studies* **13** 563-88.

[16] Greenwood R, Hinings C, Suddaby R 2002 Theorizing change: The role of professional associations in the transformation of institutionalized fields *Academy of Management Journal* **45** 58-80.

[17] Lawrence TB, Suddaby R 2006 Institutions and institutional work. In: Clegg S, Hardy C, Lawrence TB, Nord WR, editors. *The SAGE Handbook of Organization Studies* 2 ed (Sage Publications: London) p. 215-54.

[18] Dacin MT, Goodstein J, Scott WR 2002 Institutional theory and institutional change: Introduction to the special research forum *Academy of Management Journal* **45** 43-56.

[19] DiMaggio PJ 1988 Interest and agency in institutional theory. In: Zucker LG, editor. *Institutional Patterns and Organizations: Culture and Environment* (Cambridge, MA: Ballinger) p. 3-21.

[20] Buchanan D, Badham R 2008 *Power, Politics, and Organizational Change: Winning the Turf Game* (London: Sage).

[21] Jasperson JS, Carte TA, Saunders CS, Butler BS, Croes HJ, Zheng W 2002 Review: power and information technology research: a metatriangulation review *MIS Quarterly* **26** 397-459.

[22] Greenwood R, Hinings CR 1996 Understanding radical organizational change: Bringing together the old and the new institutionalism *Academy of Management Review* **21** 1022-54.

[23] Wilson EJ 2008 Hard power, soft power, smart power *The Annals of the American Academy of Political and Social Science* **616** 110-24.

[24] Dorado S 2005 Institutional entrepreneurship, partaking, and convening *Organization Studies* **26** 385-414.

[25] Wijen F, Ansari S 2007 Overcoming inaction through collective institutional entrepreneurship: Insights from Regime Theory *Organization Studies* **28** 1079-100.

[26] Tolbert PS, Zucker LG 1983 Institutional sources of change in the formal structure of organizations: The diffusion of civil service reform, 1880-1935 *Administrative Science Quarterly* **28** 22-39.

[27] Avgerou C 2000 IT and organizational change: An institutionalist perspective *Information Technology and People* **13** 234-62.

[28] Jarzabkowski P 2004 Strategy as practice: Recursiveness, adaptation, and practices-in-use *Organization Studies* **25** 529-60.

[29] Haswidi A 2007 E-tendering will improve transparency, official says *The Jakarta Post* 6 January.

[30] The Jakarta Post 2009 Corruption ingrained in procurement processes *The Jakarta Post* 5 February.

[31] Transparency International 2008 *Handbook for Curbing Corruption in Public Procurement* (Berlin: Transparency International).

[32] Jones DS 2007 Public procurement in Southeast Asia: Challenge and reform *Journal of Public Procurement* **7** 3-33.

[33] Bygstad B, Munkvold BE 2010 Exploring the role of informants in interpretive case study

research in IS *Journal of Information Technology* **26** 32-45.

[34] Myers MD, Newman M 2007 The qualitative interview in IS research: Examining the craft *Information and Organization* **17** 2-26.

[35] Yin RK 2009 *Case Study Research: Design and Methods* (California: Sage).

[36] Scott WR 2008 *Institutions and Organizations: Ideas and Interest* (Thousand Oaks, CA: Sage).

[37] Wahid F, Sein MK 2014 Steering institutionalization through institutional work: The case of an eProcurement system in Indonesian local government *Proceedings of the 47th Hawaii International Conference on System Sciences* (HICSS) 2014.

[38] Fligstein N 1997 Social skill and institutional theory *American Behavioral Scientist* **40** 397 405.

[39] Scott WR 2005 Institutional theory: Contributing to a theoretical research program. In: Smith KG, Hitt MA, editors. *Great Minds in Management: The Process of Theory Development*, (Oxford: Oxford University Press).

[40] Seldadyo H, De Haan J 2011 Is corruption really persistent? *Pacific Economic Review* **16** 192 206.

[41] Hofstede G 1991 *Culture and Organizations: Software of the Mind* (New York: McGraw-Hill).

[42] Suddaby R 2010 *Challenges for institutional theory Journal of Management Inquiry* **19** 14-20.

[43] Maguire S, Hardy C, Lawrence TB 2004 Institutional entrepreneurship in emerging fields: HIV/aids treatment advocacy in *Canada Academy of Management Journal* **47** 657-79.

[44] Lee AS, Baskerville RL 2003 Generalizing generalizability in information systems research *Information Systems Research* **14** 221-43.

[45] Heeks R, Santos R 2009 Understanding adoption of e-Government: Principals, agents and institutional dualism *iGovernment Working Paper Series* No 19

IOP Conf. Series: Materials Science and Engineering **185** (2017) 012005 doi:10.1088/1757-899X/185/1/012005

Transformation of fraud activities in procurement system in Indonesia

N Setiani[1], S N Huda[1], R Pulungan[2] and E Winarko[2]

[1] Department of Informatics, Universitas Islam Indonesia, Yogyakarta, Indonesia

[2] Department of Computer Science and Electronics, Universitas Gadjah Mada, Yogyakarta, Indonesia

E-mail: novi.setiani@uii.ac.id, sheila@uii.ac.id, pulungan@ugm.ac.id, ewinarko@ugm.ac.id

Abstract. E-procurement has been applied in Indonesia since 2008 through the adoption of Electronic Procurement System (SPSE). The purpose of the use of SPSE in the procurement process is to improve the efficiency of goods or services procurement. In the bidding process, intensive communication and data exchange between providers and organizers are urgently needed. Through SPSE, the frequency of face to face meetings between providers and the committee can be reduced. This is expected to minimize potential fraud behaviors in the goods or services procurement conducted through the tender process. There exists a transformation of activities in the procurement through SPSE from the tender process that was previously done manually. In this paper, we analyze this transformation between manual procurement process and SPSE-based procurement process. The result of the analysis is exploited for investigating the possibility of fraud behavior transformations in every phase of e-procurement activities.

1. Introduction

E-procurement has been applied in Indonesia since 2008 through the adoption of Electronic Procurement System (SPSE). According to LPSE smart report in 2015, the implementation of e-procurement was pioneered in 2008 by the Electronic Procurement Services (LPSE) of the Ministry of Finance with 2 tenders, LPSE of the Ministry of Education and Culture with 3 tenders, LPSE Denpasar with 2 tenders, LPSE Yogyakarta with 11 tenders, LPSE Special Region Yogyakarta with 2 tenders, LPSE Gorontalo Province with 6 tenders, LPSE West Java province with 3 tenders, LPSE East Java Province with 2 tenders, LPSE Riau Islands province with 1 tender, and LPSE West Sumatra province with 1 tender. Thus, there were in total 33 tenders and all of them are completed. These LPSEs rendered the tender processes more efficient by more than 6.6 billion Rupiahs, which is equivalent to 15.41% of the total budget ceiling. Each government agency has been required by Presidential Decree No. 54 of 2010 to use SPSE in each of their procurement above 200 million Rupiahs starting from 2012. The current number of SPSE users has reached 635 LPSEs distributed throughout Indonesia.

On the other hand, in the implementation of e-procurement, there exist some obstacles such as the availability of infrastructure in Indonesia and the capability of human resources in adopting SPSE. SPSE requires Internet connection for the entire procurement process, but the infrastructure for Internet services in several regions in Indonesia has not been available. Not only Internet services, limited supply of electricity and frequent rolling blackouts also may obstruct the implementation of e-procurement. The problems of human resource aspect are related with the limited knowledge of the stakeholders, both the tender committee and the vendor companies participating in the tender process through e-procurement [1].

Goods/services procurement consists of planning process, product design process, provider selection (tender) process, execution of the contract process, and warranty and audit process. Procurement planning process is conducted by government agencies through the budget process. Selection of providers can be done through auction, direct procurement, direct appointment and e-purchasing. In the implementation phase of the contract, vendors realize the procurement of goods/services that has been promised. During the warranty period, vendors provide maintenance services and follow the audit process of the implementation of the of goods/services.

The purpose of the use of SPSE in procurement process is to improve the efficiency of the tendered goods/services procurement. In the bidding process, communication and intensive data exchange between providers and organizers are necessary. Through SPSE, the frequency of face to face meetings between providers and the committee can be reduced. This is expected to minimize potentials for fraud behaviors in the procurement of goods/services conducted through a tender process.

There is a transformation in the activities of goods/services tenders through SPSE from the tender processes of goods/services that were previously done manually. In this paper, we analyze the transformation between manual procurement processes and the auction processes through SPSE. The results of the study will be used to discuss the impact of the adoption of e-procurement for the activity that occurs.

2. E-procurement

The government's public procurement process consists of several phases and every phase is risky with corruption. Mathecak [2] identified three main phases in the procurement process: procurement planning and budgeting, procurement solicitation, and contract award and performance. Szymanski [3] instead proposed five phases of a procurement process: procurement planning and requirement assessment, product design and documentation, tender process, contract award and implementation, and accounting and auditing. The risks of corruption arise from the lack of transparency, limited access to information, and the lack of accountability and control at each phase.

Public e-procurement is defined as the use of information and communication technology (ICT) such as Internet or web based system by the government to carry out the procurement associated with bidders to acquire a product, employment, and other services needed by the public sector [4, 5]. E-procurement is defined as well as inter-organizational information system, which automates the entire part of the procurement process to improve efficiency, quality, and transparency in government procurement [6].

Lewis-Faupel et al. [7] conducted a study on the implementation of e-procurement in the two large developing countries, Indonesia and India. They arrived at a conclusion that there was not sufficient evidence to conclude that the implementation of e-procurement can reduce the amount of fees to be paid by the government. However, they found also a conclusion that the implementation of e-procurement in Indonesia has an impact on the reduction of delays in the completion of projects. On the other hand, the implementation of e-procurement also provides an opportunity for more variations in vendors/suppliers of goods and services that participate in the tender and the winning bidders tend to be more varied than before and have a better quality (measured from reduced delays in the completion of projects in Indonesia or the quality of jobs in India).

Neupane et al. [8] in their study discussed the direct impact of the implementation of e-procurement to the reduction of corruption. Identified factors that are prone to corrupt practices are project planning, product design and documentation, tendering processes, contract enforcement, and accounting and auditing. Research result indicated that the ability of e-procurement in terms of anti-corruption through automation and audit capabilities has increased transparency and accountability of the auction process by the government.

3. Research Method

We use qualitative approach for conducting our research. Qualitative methods are a set of interpretative techniques that attempt to explain and extract meaning from a naturally occurring phenomenon in the social world [9]. We conduct our research in four steps as follows:

(i) Literature review. At this steps, we perform a study of the various sources of literature in the procurement process, especially in Indonesia. The sources of literature include journals, papers, books, and credible websites such as lkpp.go.id, and bpkp.go.id. Topics studied covert procurement and SPSE (e-procurement in Indonesia).

(ii) Data Collecting. To collect facts about procurement and implementation of SPSE in Indonesia, we carry out interviews with LKPP (Policy Institute for Government Procurement of Goods/Services), experienced vendors, and LPSE manager at several institutions such as state universities, and provincial and local government agencies.

(iii) Data Analysis. We use open coding analysis for mapping the result of the data collected and the literature review to build a model describing the transformation of procurement activities and potential fraud behaviors in SPSE implementation.

(iv) Conclusion making. And the last, we make conclusion of our research based on the result of data analysis process.

4. Adoption of SPSE as Indonesian e-procurement system

SPSE is Indonesia's e-procurement system built by LKPP. Now, SPSE has reached version 4, although not all government institutions have used SPSE version 4. Many government institutions still use SPSE version 3. In this research, we limit the area of analysis to SPSE version 3 and 4 because lack of documentation of SPSE version 1 and 2. In version 3, the e-procurement system emphasizes the exchange of online documents, such as the announcement of procurement plan through the web, uploading procurement documents by tender committee, so it can be downloaded by all interested prospective providers, as well as uploading bidding documents by potential providers so the procurement committee can download them. Discussions between organizers and providers before uploading providers' bidding documents are recorded in the system, so the conversation can be viewed and audited. The evaluation process is carried out offline, and then the evaluation result is uploaded to the system. There may be potential provider's errors when making quotation, thus, the committee needs to recalculate (arithmetic correction). The result of the recalculation will also be uploaded to the system. Determining the winner of the tender, however, is carried out manually, and then the result is entered into the system. SPSE version 3 therefore aims at transparency, because most of the procurement committee activities are recorded in the system.

SPSE version 4, on the other hand, has been semi-automatic. Procurement documents are generated by the system, so the tender committee only needs to input data to the system. Official reports can also be generated by the system so no need to create them manually. Uploading price by potential providers is also using a web page that resembles Microsoft Excel spreadsheet with automatic calculation, so there is no required arithmetic correction from the committee. For the auction by knockout, a potential provider with the lowest bid automatically wins in the SPSE version 4. SPSE version 4 has been used in about 100 LPSEs of local government level. The transformation of procurement activities is presented in table 1.

Table 1. Transformation of Procurement Activities.

No	Activity	Before SPSE	SPSE version 3	SPSE version 4
1	Tender announcement	Information board in the committee office.	Website.	Website.
2	Procurement document	Written by tender committee and saved in hardcopy format.	Written in PDF document and uploaded to the website by tender committee.	Submitted in website form by the tender committee and generated by the system.
3	Vendor quotation	Submitted by vendor to the office in hardcopy format.	Uploaded by vendor to the website in PDF format.	Submitted in website form by the vendor and generated by the system.
4	Aanwijzing	Vendor and tender committee meet in the committee office.	Vendor and tender committee join the forum in the website.	Vendor and tender committee join the forum in the website.
5	Evaluation	Tender committee evaluates based on vendor quotation documents.	Tender committee evaluates based on vendor quotation documents.	Semi-automated by the system, especially for the price components

SPSE is developed by LKPP to run by LPSE (Electronic Procurement Services) as institutions responsible for running SPSE. At the beginning of 2008 when SPSE was first applied, there are only 10 LPSEs. Now the total number has reached 635 LPSEs The growing process of LPSEs is a process that takes time and effort. It mainly depends on the political will of the local governments and on the awareness of the need for SPSE application. With a corrupt culture in many previous procurement, initial resistance to the application of SPSE is strong. Until 2010, the number of LPSEs only reached 137 LPSEs in 28 provinces. With the Presidential Decree no. 54 of 2010, which states that the e-procurement should be applied starting in 2012, the number of LPSEs in Indonesia increased considerably. In 2011, the number of LPSEs has reached 315 LPSEs that cover 31 provinces. In 2012, the number of LPSEs reached 543 LPSEs which cover 33 provinces.

To get willingness of each region to implement SPSE has not been easy. If there is a willingness of the leadership, it is easy for the region to adopt SPSE. But if there is no willingness, especially if there are many local authorities that play in procurement, coercion is needed as in the presidential decree. In addition to the problem of willingness of local government leaders, another main problem is the budget. Many areas refuse to implement SPSE by reason of the lack of budget. Another problem is the lack of personnel to form LPSE. Ad hoc LPSEs are not uncommon, where personnel in those LPSEs has main duty in her base institutions and her duty in the LPSEs is only a side job. Some of the obstacles cause LPSE structure between one region and another in Indonesia is not under the same institution. LPSE placement is very diverse, for example, in the Development Administration, Bappeda, Diskominfo, or UPT ULP. Resistance and willingness factors affect the commitment of personnel and the committee in conducting procurement activities through SPSE. Potential fraud behaviors, such as subjectivity from the committee on provider assessment, have higher possibility to arise in the ULP with lower willingness and higher resistance.

The next obstacle is the problem of infrastructure: e-procurement focuses on the use of the

Internet for the exchange of data and information, while there are many areas in Indonesia with inadequate Internet access. In fact, SPSE is decentralized in each LPSE server. Hence, especially the eastern parts of Indonesia face obstacles in building the necessary infrastructure for LPSE server. SPSE implementation, however, has a positive impact, where the local governments are forced to build Internet infrastructure in cooperation with major ISPs so they can build LPSE servers. Nowadays there are 635 LPSE servers running decentralized SPSE. The decentralized servers will later cause some problems because of the condition of the servers that varies between each LPSE. Nonetheless, the initial goal of decentralized LPSE servers is to increase the reliability of the overall procurement. For example, in a centralized server, when the server is down, then the entire electronic procurements in Indonesia is hampered. In decentralized servers, if one LPSE server goes down, then it only hampers the procurement at the respective LPSE.

On the other hand, the infrastructure problem is also felt by the providers, where not all providers have good Internet access. With decentralized LPSE servers, in each LPSE a bidding room is available to serve providers to access SPSE via local network (LAN). Potential providers who have difficulty accessing SPSE via Internet may come and use the LPSE bidding room to access SPSE, such as uploading the bidding documents. The varying condition of infrastructure and server decentralization also affects the level of system security in the LPSEs. Although LKPP has defined a system security standard based on ISO 27001, it is not well implemented in a number of LPSEs.

5. Transformation of procurement activities through SPSE

As mentioned above, Szymanski [3] has identified five phases of procurement process, *i.e.*, procurement planning and needs assessment, product design and documentation, tender process, contract award and implementation, and accounting and auditing. The planning stage of procurement lies fully in the ministries, agencies, regional work units or other institutions, namely those using the national budget (APBN) or regional budget (APBD). In this planning phase, there is still potential for cheating, for example with a mark-up during the preparation of the budget by raising the value of the budget to unreasonable price for personal profit. Now the General Procurement Plan (Rencana Umum Pengadaan/RUP) has been made public, though not detailed, and administered using SIRUP (Sistem Informasi Rencana Umum Pengadaan/Information System for General Procurement Plan). LKPP also has plans to integrate SIRUP, SPSE, SIKAP (Sistem Informasi Kinerja Penyedia/Information System for Provider's Performance), contract and financial management applications. The focus for the development in this year (2016) is to build SIKAP and integrate it with other existing systems.

The implementation of e-procurement has a positive influence in transparency. Prior to the implementation of e-procurement, procurement in Indonesia were still fraught with corrupt practices, particularly in the phase of the tender process. This started from the announcement of the tender, which was usually done using notice boards and newspapers. In remote areas, cases of tender announcement removals from notice boards—with the aim that other potential providers would not know the procurement information—were often encountered. Announcements through newspapers could be sabotaged by big providers by means of buying up all the newspapers immediately after printing.

After the announcement phase of the procurement plan, potential providers who are interested in participating the tender need to take the procurement documents in the government institution that organizes the tender. This stage in the past, before using e-procurement, was also vulnerable to fraud. The use of thugs to stand by in the institution office by cheating big providers often happened so that other providers cannot take the procurement documents.

The next stage is aanwijzing; that is the process of auction explanation. The tender committee accompanied by a technical team explains the procedures for the auction and the technical specifications of the procurement. At this stage, there was a direct face to face between the tender committee and the potential providers. Direct face to face could trigger the initiative of those who wanted to commit fraud. Not only between committee members and providers, fellow providers could conduct discussions and project distribution. As for the committee and the providers, for instance, if there is an addendum or specification changes that must be made public, the tender committee that wanted a certain provider to win could cheat by informing the addendum to the preferred provider first so that such provider could have more time to prepare the bidding documents in accordance with the addendum.

After that, providers must enter the bidding documents. At this time, the use of thugs (especially in rural areas) reoccurs. Some providers are intimidated by people from big providers so they cannot submit their bidding documents.

The entered bidding documents are then evaluated by the procurement committee to determine the winner of the auction. In general, the evaluation consists of evaluation of the qualifications, technical, pricing, and administration. Potential frauds are on qualification and technical evaluations because they involve assessments that rely on the procurement committee. The committee could subjectively elevate the value of the technical evaluation of some preferred providers. This subjectivity is usually based on personal ties, personal background, as well as elements of fraud.

After the implementation of e-procurement, some frauds can be averted, although e-procurement cannot solve all problems. The announcement of the procurement plan is done through each LPSE's web, so there is no potential for taking off and buying all newspapers announcing the plan. Procurement documents can be downloaded online, so there is no potential to play the role of thugs. Similarly, the bidding documents can be uploaded online by potential providers, so there is no potential for intimidation by thugs. However, there are still other methods to commit fraudulent behaviors using e-procurement as described in table 2.

As for the subjectivity factor, for example relating to personal ties during the technical and qualification evaluation by the committee, it is beyond SPSE scope. Because the qualification evaluation and technical assessment are based on human judgement, it is still possible that assessments are not appropriate, and favor one provider over the others. It is admitted by a provider who becomes our interviewee that the assessment of qualification and technical evaluation is still playable. Despite the stage of proof of qualification and assessment guidelines rubric, as human beings, elements of subjectivity remain. Automatizing the process of qualification and technical evaluation is difficult. On the other hand, the way a provider presents itself can be one way for the tender committee to assess the its qualification. The element of subjectivity is unavoidable, can even be regarded as art in the tender. This has to be guarded so that assessments are more objective, for example by strengthening the assessment rubric or by more thorough audit.

6. Concluding remarks

SPSE version 4 emphasizes on the semi automation of generating procurement documents and pricing input, while SPSE version 3 emphasizes on the recording of procurement activities to provide transparency in procurement process. The adoption of SPSE has made some changes in the procurement activities. The changes eliminate several potential frauds that exist in manual procurement systems especially in two activities, namely tender announcement and procurement document dissemination. However, subjectivity factor in technical and qualification assessment still exists. There is also a need to integrate existing systems to link every phase in the procurement process.

Table 2. Potential Fraudulent Behaviors during Procurement Activities.

No	Activity	Before SPSE	SPSE version 3	SPSE version 4
1	Tender announcement	A vendor obstructs other vendors from getting procurement announcement, for example by revoking the announcement from the information board or by buying all newspapers containing tender information.	N/A	N/A
2	Procurement document dissemination	A vendor obstructs other vendors from getting procurement documents, for example by using thugs to stand by in the committee office.	N/A	N/A
3	Vendor quotation	A vendor intimidates other vendors so they cannot submit bidding/quotation documents.	Deadline of quotation is not clear. Bandwidth setting or DDOS attack.	Deadline of quotation is not clear. Bandwidth setting or DDOS attack.
4	Aanwijzing	The committee and the vendor compromise about the addendum of tender specifications.	An internal (ex-internal) admin of SPSE edits or deletes quotation documents of some vendors in the Aanwijzing phase.	N/A
5	Evaluation	The committee assesses quotation documents based on the deal with the vendor.	The committee submits assessment based on their preferences or deal with the vendor. Evaluation process is conducted by an external person who infiltrates the system [10].	The committee submits assessments based on their preferences or deal with the vendor.

Acknowledgments

The authors would like to thank the Ministry of Research, Technology and Higher Education of Indonesia for funding this research.

References

[1] Purwanto A, Ibty I, Rofikah N and Indroyono P 2008 E-procurement in Indonesia (development services procurement electronic government) *Kemitraan Partnership dan LPSE Nasional*

[2] Matechak J P 2002 Fighting corruption in public procurement *Center for International Private Enterprise CIPE, Feature Service Articles* **12** 1047–55

[3] Szymanski S 2007 How to fight corruption effectively in public procurement in SEE countries *Concept Paper* URL http://www.oecd.org/dataoecd/24/28/39637617.pdf

[4] Davila A, Gupta M and Palmer R 2003 Moving procurement systems to the Internet: the adoption and use of e-procurement technology models *European Management Journal* **21** 11–23

[5] Leipold K, Holloway F, Klemow J and Vaidya K 2008 The world bank e-procurement for the selection of consultants: challenges and lessons learned *International Handbook of Public Procurement* ed Thai K V (Auerbach Publications) pp 512–27

[6] Vaidya K and Hyde M 2011 Inter-organisational information systems assimilation: an empirical evaluation in light of the diffusion of innovation theory *International Journal of Business Information Systems* **7** 247–68

[7] Lewis-Faupel S, Neggers Y, Olken B A and Pande R 2016 Can electronic procurement improve infrastructure provision? Evidence from public works in India and Indonesia *American Economic Journal: Economic Policy* **8** 258–83

[8] Neupane A, Soar J, Vaidya K and Yong J 2012 Role of public e-procurement technology to reduce corruption in government procurement *5th International Public Procurement Conference (IPPC5)* pp 304–34

[9] Cooper D and Schindler P 2008 *Business Research Methods* The McGraw-Hill/Irwin series operations and decisions sciences (McGraw-Hill)

[10] Arumsari, Totok, Iswahyudi, Mucharor and Akib 2014 Audit Atas Pelaksanaan Lelang Secara Elektronik dalam Pengadaan Barang dan Jasa Pemerintah *Badan Pengawasan Keuangan dan Pembangunan* URL http://www.bpkp.go.id/investigasi/berita/read/13521/0/AUDIT-ATAS-PELAKSANAAN-LELANG-SECARA-ELEKTRONIK-DALAM-PENGADAAN-BARANG-DAN-JASA-PEMERINTAH.bpkp

Cross-country Analysis of ICT and Education Indicators: An Exploratory Study

Ahmad R Pratama[1,2]

[1] Department of Informatics, Universitas Islam Indonesia, Sleman, DIY 55584 Indonesia
[2] Department of Technology and Society, Stony Brook University, NY 11794 USA

E-mail: ahmad.rafie@uii.ac.id; ahmad.pratama@stonybrook.edu

Abstract. This paper explores the relationship between world ICT and education indicators by using the latest available data from World Bank and UNESCO in range of 2011-2014 with the help of different exploratory methods such as principal component analysis (PCA), factor analysis (FA), cluster analysis, and ordinary least square (OLS) regression. After dealing with all missing values, 119 countries were included in the final dataset. The findings show that most ICT and education indicators are highly associated with income of the respective country and therefore confirm the existence of digital divide in ICT utilization and participation gap in education between rich and poor countries. It also indicates that digital divide and participation gap is highly associated with each other. Finally, the findings also support the idea of reverse causality between ICT and education; higher participation rate in education increases technology utilization, which in turn helps promote better outcomes of education.

1. Introduction

It is a fact that technology, particularly information and communication technology (ICT), has been growing rapidly within the past decade. While it exists in different forms from telephone, either landline and mobile, to the Internet, they all are growing at an exponential rate and have been impacting the shapes of human lives, for better or for worse. It connects people in different parts of the world, breaking the barriers of time and distance. Nevertheless, despite some negative side effects one might argue about it, it does open numerous possibilities to improve the quality of human lives. One of them is by improving the quality of education.

In this paper, we try to explore more about the relationship between ICT and education by using worldwide indicators, summarize several indicators to see if there is something in common or the point out important differences among them, examine different level of education in term of technology utilization, and see what effect technology utilization has to education outcomes. Furthermore, we will also examine if income has something to do in the relationship between ICT and education in many countries.

2. Conceptual Background

Two decades ago, long before the era of Google, Wikipedia, or Facebook, Kuntz suggested that the Internet promotes literacy [1]. Five years later, another scholar argued that ICT is an option to close the gap of educational inequalities [2]. Ever since, the notion that ICT helps improve education has been widely accepted with the fact of the rise of many e-learning platforms, one of which is well known as massive, open, online course (MOOC) as well as the popularity of Educational Technology as a new research area. The more recent trend also includes mobile devices where we no longer use them only to make phone calls or to send text messages with the introduction of smartphones. Nowadays, people can play games, socialize with other people across the globe, and of course educate themselves by using their own mobile devices. Many have declared their belief in the opportunities and advantages of mobile learning [3,4].

Interestingly, it seems that the relation between ICT and Education is not a one-way causality. Some other researchers argued that technology utilization is also driven by social factors, one of which is education [5, 6]. Using data from 2008-2009, the same researchers showed that tertiary enrollment rate is one of the most important factor driving technology utilization [7]. In other words, the higher tertiary enrollment rate in a country, the more people in that country will use ICT. These findings suggest that there is actually reverse causality between ICT and education and this relationship is worth to explore more, preferably with the most recent data available. Furthermore, it is also interesting to see how different the trend and the relationship in poor and rich countries is, given that digital divide has been widely accepted as a world problem for the past decade [8].

3. Data and Methodology

3.1. Variables of Interest

There are four indicators of ICT and five indicators of education with one additional indicator serving as a control variable used in the analyses. The indicators for ICT are; 1) Internet Users (per 100 people), 2) Fixed/Wired Broadband Subscriptions (per 100 people), 3) Mobile Cellular Subscriptions (per 100 people), and 4) Secure Internet Servers (per 1 million people). These indicators are selected since they represent ICT penetration and are commonly used by many world organizations such as World Bank, OECD, and ITU as key ICT indicators. Furthermore, one particular indicator (i.e. Secure Internet Servers) also represents ICT infrastructure strength of the country as it shows producer point of views rather than consumer's as the other three do.

Meanwhile, the education indicators are divided in two main categories; participation in education, represented by 1) Primary Enrollment Rate (%), 2) Secondary Enrollment Rate (%), 3) Tertiary Enrollment Rate (%), and the outcomes of education, represented by 4) Literacy Rate (%), and 5) Scientific and Technical Journal Articles Published. For the participation in education, it is quite obvious to include all three levels of education while the two different indicators for outcomes are used to accommodate this difference in education level. The first indicator is more associated with primary and secondary education while the last one is used to measure the indicator of tertiary education. Furthermore, these two indicators are also selected as they both can be seen as direct outcomes of education, thus minimizing the potential effect of external factors that are not included in this study.

Lastly, GDP (per capita) is used as a measure of income serving as a control variable in this study. All data were taken mainly from the World Bank using the average value of any available data within 2011-2014. Some missing values for Literacy Rate were filled by using additional data from UNESCO for the same time range. After dealing with all missing values, 119 countries were included in the final dataset.

3.2. Analysis Methods

Principal component analysis (PCA) was used to answer the first research question about the pattern and directions of these ICT and education indicators. The same PCA in addition to factor analysis (FA) were then used to summarize the data. In both PCA and FA, correlation matrix was used as opposed to covariance matrix due to metric differences between variables. Furthermore, to compare the differences between poor and rich countries, PCA was repeated for two subsets, each representing poor countries (i.e. GDP per capita less than $1,047) and rich countries (i.e. GDP per capita over $12,000) in addition to cluster analysis (CA) to see whether or not poor and rich countries fall under different clusters. Lastly, ordinary least squares (OLS) regression is used to find the impact of educational level (primary, secondary, and tertiary) on technology utilization as well as the impact of ICT indicators on educational output (journal articles and literacy rate).

As 6 out of 10 variables were either highly skewed (i.e. GDP per capita, secure internet servers, and journal articles), moderately skewed (i.e. broadband and tertiary enrollment), or negatively skewed (i.e. literacy rate), log transformation was done to deal with highly skewed distributions and square root transformation was performed on the moderately skewed variables. Meanwhile, dealing with negatively skewed distribution involved three steps transformation where a reflection was done at the first step (i.e. by subtracting it from the maximum possible value) before log transformation was applied. The final step was to reflect back the log transformed variable so it resembles the original values (i.e. by subtracting it from the maximum value of the log transformed variable).

Unfortunately, even after the transformation, the dataset still failed to pass multivariate normality test. Therefore, the analyses were then conducted with both the original as well as the transformed data and so that the results between two of them can be compared. Apparently, there was no significant difference when the analyses were done with either the original or the transformed data. However, the transformed data provide much better results in PCA, FA, and CA. The original dataset was used in OLS Regression and log transformation will be done separately in each regression model as necessary. In term of outliers, there are three multivariate outliers detected for the transformed data (i.e. Cuba, Sudan, and Turkmenistan). However, these outliers do not affect the findings since we got very similar results whether or not the outliers are excluded from all analyses. Meanwhile for the rich countries and poor countries subsets, the transformed data did pass multivariate normality test. No outlier was detected for the rich countries and one non-influential outlier (i.e. Chad) was detected for the poor countries.

4. Findings and Discussions

The PCA loadings show that the first principal component of this dataset is associated with nine out of ten variables with evenly distributed proportion, meanings those nine variables are associated to each other very well. Furthermore, the biplots as can be seen in Figure 1 show that all four ICT indicators and four out of five education indicators go to the same direction with income. Only primary enrollment ratio goes to quite the opposite direction. This suggests that the level of technology utilization as well as the participation and the outcome of education are heavily affected by income. Richer countries tend to have more technology utilization and more participation in education as well as better outcome.

When it comes to ICT indicators, it turned out that all indicators go to relatively the same main direction. However, it is important to note that mobile subscription number has slightly different deviation compared to the other three indicators due to the fact that not all people use their mobile devices to access Internet. While it is true that Internet access from mobile devices such as smartphones and tablets has become more and more popular lately, some people are still using regular cell phones that are mainly used merely to make phone calls and to send text messages.

As for education indicators, the biplot emphasizes the previous results that primary enrollment has quite large deviation compared to all other indicators. One possible explanation is that because most countries mandate primary education for their citizen while it is not the case for secondary and tertiary education. That being said, changes in primary enrollment ratio does not really affect the education outcomes directly, just like what the biplots show that it does not go to the same direction with other education indicators. However, it is important to note that significant drop in primary enrollment rate might affect secondary and tertiary enrollment rate and therefore it may have an indirect effect on the outcomes of education.

In terms of data summarization, PCA can explain more of the total variance but also produces more outliers, all of them are either poor countries or rich countries with extreme low or high values in two or more variables. While FA cannot explain as much variance as PCA using the same number of factors, it produces only one outlier. Similar case applies to education indicators where FA can explain 5% less variance than PCA, but with the use of varimax rotation, it has the advantage of having only one outlier, compared to 11 outliers with PCA, and the inclusion of significant countries such as China. Considering China is accounted for roughly 20% of world population, it is not a wise decision to simply exclude it from the dataset. Therefore, in both occasions, FA is more preferable than PCA to summarize this dataset. As for ICT indicators, PCA wins by a margin since it can summarize more variance with fewer outlier than FA.

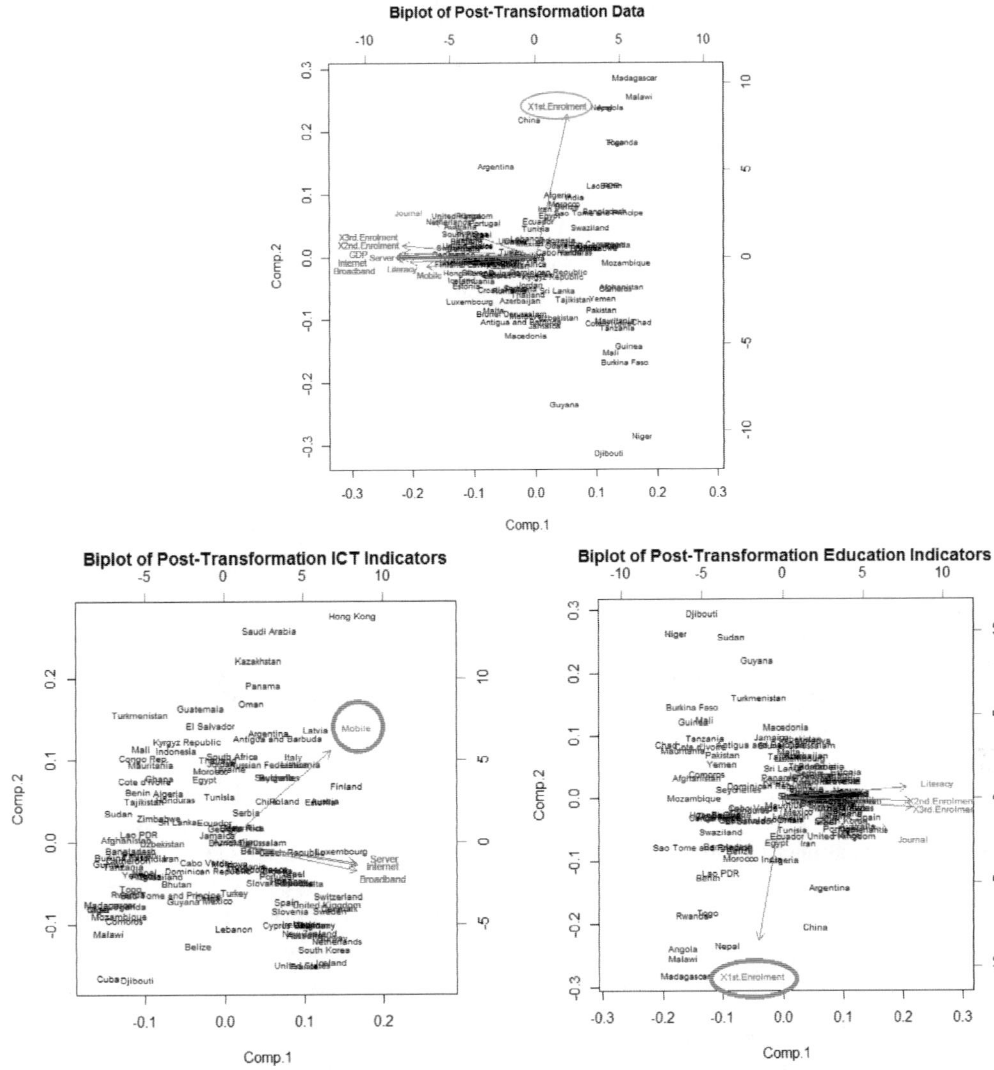

Figure 1. PCA biplot for a) all indicators, b) ICT indicators, c) education indicators.

The findings also confirm that digital divide and participation gap do exist to date since poor countries and rich countries tend to fall under different clusters no matter what clustering method is used to the dataset. The dendrogram also identified some countries with extreme values in one or more variables such as United States that produces the most journal articles worldwide and Hong Kong that has the highest mobile subscription worldwide. Both countries lead by a very large margin on each variable respectively, making them very distinctive. The PCA biplots (Figure 2) also show that in poor countries, income is associated more with participation in education while in rich countries it is associated more with the Internet (i.e. Internet users, broadband subscriptions, and Internet servers). It is also interesting to see that in poor countries, broadband and mobile subscriptions tend to be a substitute since they are fetching to the opposite direction from each other. Meanwhile, if we measure the outcome of education by using the number of journal articles published by scholars of a country, it is associated with the participation in secondary and tertiary education, but only for rich countries. It seems to have nothing to do with the participation in education for poor countries. Again, this confirms that digital divide and participation gap is associated with each other.

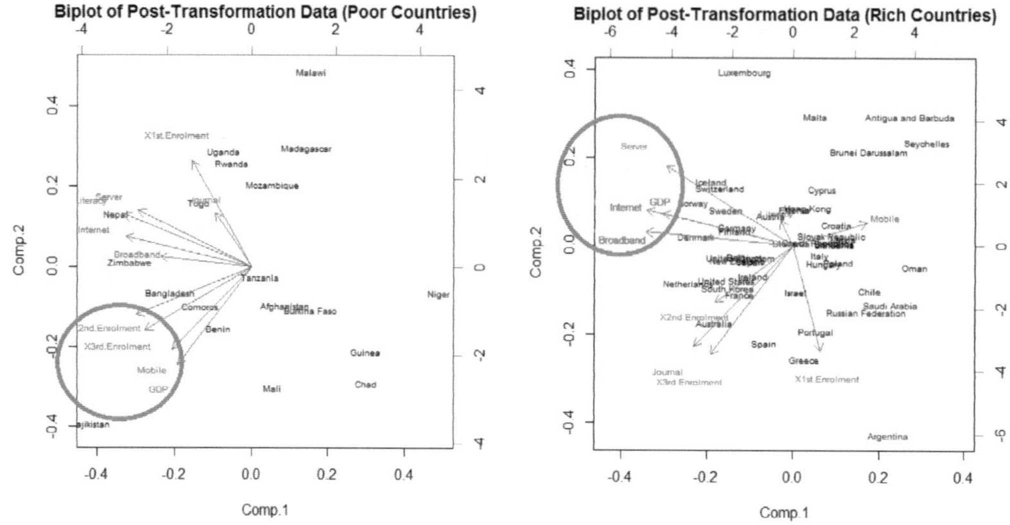

Figure 2. PCA biplot for all indicators in poor countries (left) and rich countries (right).

Finally, the OLS regression results in Table 1 show that the Internet does promote literacy. It confirms the theory still holds true even after two decades. Using stepwise deletion method by putting all ICT indicators and GDP per capita as independent variables, we can see that for every one more person uses the Internet per 100 people and by holding GDP per capita constant, the literacy rate tends to increase by 23%. Furthermore, when we measure the outcomes of education based on the number of journal and scientific articles published by a country, it tends to increase by 22% for every one additional broadband Internet subscription per 100 people, holding all other predictors constant.

Table 1. OLS regression estimates for education outcomes

	Literacy Rate (^)	**Journal Articles** (^)
GDP per capita	-.061 ***	-
Secure Internet Server	-	-1.4×10^{-3} **
Internet Users	.23 ***	-
Broadband Subscribers	-	.22 ***
Mobile Subscribers	-	-
Constant	4.26 ***	3.44 ***
Adjusted R^2	.655	.450
Highest VIF	4.31	2.80
Number of Cases	119	119

Note: Unstandardized coefficient, two tailed-test
* p < .05 ** p < .01 *** p < .001 (^) log transformed

Furthermore, as shown in Table 2, secondary and tertiary education are indeed highly significant predictors for technology utilization. The higher secondary and/or tertiary enrollment ratio in a country, the more people in that country use ICT. It is interesting to see that secondary education plays more important role than tertiary education for mobile devices penetration. It suggests that the use of mobile devices is highly associated with adolescents more than with adults, just like what was suggested in the literature [8]. It is also worth to note that many researchers found excessive and problematic use of mobile devices among adolescents [9, 10]. Meanwhile, tertiary education is highly associated with broadband Internet connections more than any other education level. This can be explained by the argument that higher education utilizes broadband Internet connection much more than in any other education level through e-Learning platforms and to support research activities as described by the literatures [11]. Considering this kind of activities require large bandwidth and stable connection, it is understandable why it is related more with broadband Internet connection rather than any kind of Internet connection.

Table 2. OLS regression estimates for technology utilization

	Secure Internet Server (^)	**Internet Users** (^)	**Broadband Subscriptions** (^)	**Mobile Subscriptions** (^)
GDP per capita	5.33×10^{-5} ***	5.89×10^{-4} ***	3.21×10^{-4} ***	.10 ***
Secondary Enrollment	3.52×10^{-2} ***	.34 ***	-	.41 ***
Tertiary Enrollment	2.47×10^{-2} ***	.31 ***	.21 ***	-
Constant	-1.31 **	-7.41 *	-1.76 *	1.88 ***
Adjusted R^2	.743	.849	.830	.487
Highest VIF	2.52	2.52	1.46	2.55
Number of Cases	119	119	119	119

Note: Unstandardized coefficient, two tailed-test
* p < .05 ** p < .01 *** p < .001 (^) log transformed

The results of OLS regression also support the idea of reverse causality between ICT and education. For example, higher tertiary enrollment rate helps promote more broadband Internet subscriptions in a country. On the other hand, having higher broadband Internet subscriptions rate helps foster the number of journal and scientific articles published by the scholars in the respective countries. In other words, we can see the pattern that participation in education promotes more technology utilization, while technology utilization helps improve the outcomes of education.

5. Conclusion

In this paper, several exploration methods are used to confirm that most ICT and education indicators of a country are highly associated with each other as well as with income of the respective country. It is also confirmed that the digital divide in ICT access and participation gap in education are still around and remain as a worldwide problem to solve. Moreover, these two problems are highly associated to each other with income playing a big role in it. The data also suggest that in poor countries, income is associated more with participation in education while in rich countries it is associated more with technology utilization. The findings also support the idea of reverse causality in ICT and education; higher participation rate, especially in secondary and tertiary education level, increases technology utilization while higher technology utilization of a country helps promote better outcomes of education. Combined with the previous findings when income is used as a control variable, it means that people in poor countries are still one step behind when it comes to betterment of education outcomes. Furthermore, it is also interesting to note that fix broadband subscriptions are more associated with tertiary enrollment while mobile subscriptions are more associated with secondary enrolment. It indicates some differences in needs and/or preferences between adults and adolescent. Finally, both FA, especially with varimax rotation, and PCA have their own benefit and drawback when used to summarize this cross-country dataset consolidated from the 2011-2014 World Bank and UNESCO data. Careful consideration must be taken when choosing either one as a data reduction technique for this dataset.

References

[1] Kuntz P S 1995 *Swahili-L: The Internet promotes literacy* (Computers & Education) **243** 177-81.
[2] Kelley-Salinas G 2000 *Different educational inequalities: ICT an option to close the gaps* (Centre for Educational Research and Innovation - National Center on Adult Literacy Learning to bridge the digital divide Education and skills)
[3] Ally M (Ed.) 2009 *Mobile learning: Transforming the delivery of education and training* (Athabasca University Press)
[4] Goksu I and Atici B 2013 *Need for Mobile Learning: Technologies and Opportunities.* (Procedia-Social and Behavioural Sciences) **103** 685-69.
[5] Pick J B and Azari R 2011 *A Global Model of Technological Utilization Based on Governmental Business-Investment Social and Economic Factors* (Journal of Management Information Systems) **281** 49-84.
[6] James J 2012 *The ICT Development Index and the digital divide: How are they related?* (Technological Forecasting and Social Change) **793** 587-94
[7] Pick J B and Nishida T 2015 *Digital divides in the world and its regions: A spatial and multivariate analysis of technological utilization* (Technological Forecasting and Social Change) **91** 1-17
[8] Pratama A R and Al-Shaikh M 2012 *Relation and Growth of Internet Penetration Rate with Human Development Level from 2000 to 2010* (Communications of the IBIMA).
[9] Katz J E (Ed.) 2002 *Machines that become us: The social context of personal communication technology* (Transaction Publishers)
[10] Ha J H, Chin B, Park D H, Ryu S H and Yu J 2008 Characteristics of excessive cellular phone use in Korean adolescents *CyberPsychology & Behavior* **11(6)** 783-4.
[11] Chóliz M 2012 Mobile-phone addiction in adolescence: the test of mobile phone dependence (TMD) *Prog Health Sci* **2(1)** 33-44.
[12] Conole G and Alevizou P 2010 *A literature reviews of the use of Web 2.0 tools in Higher Education.* (A report commissioned by the Higher Education Academy)

The Use of ICT to Support Perpetual Undergraduate Students

N.W. Rahayu, S.N. Huda

Universitas Islam Indonesia, Yogyakarta, Indonesia

nnur@uii.ac.id, sheila@uii.ac.id

Abstract. Perpetual students are problematic, both for campus and for themselves. Inefficient student management could lead to bad lecturer: students' ratio and cause complicated problem for students and parents. This paper describes ICT used by an Informatics department of a big private university in Indonesia to help 203 perpetual undergraduate students finishing their study in a short time. Lengths of study are varying from 7 until 15 years and most of them suffered from some compulsory credits and mandatory internship project. We observed wide-range of ICT used for data management and communication during the beginning, middle and the end of periods. Success rate of finding perpetual students and producing graduates are almost 70%, but this percentage could be higher if we maximize ICT use. Information sharing type, social media, privacy and patience become important issues related with the use of ICT.

Keyword: perpetual student, ICT use, university student, social media

1. Introduction

We all hope to be able to be perpetual learners. However, being a perpetual student is a different matter. In Indonesia, a senior student in undergraduate level who does not graduate after 5 years is often referred to as the perpetual student. Although in general the various universities in Indonesia have a policy of maximum 7-year study period for undergraduate level, but in reality there are found many students with more than 7 years study period through various dispensations from the university. In fact, most prolonged students are not active students, which did not pay semester tuition and also did not follow the lecture. Some of prolonged students always pay semester tuition but did not take any courses. There are also prolonged students whom status is working on final project but in reality they did not make any progress.

This problem arises when the study was not completed within the expected time period (4 years) in which the college curriculum in Indonesia is designed for a period of 4 years. By not achieving timely graduation, the study duration becomes longer, and some failures as well as the attractiveness of activities in addition to learning, such as part-time jobs, may lead students to become perpetual student. Based on [1], there are three negative sides of the phenomenon to prolonged study period. First, the low qualification of the student entries. Students who lack academic skills, which in the past are not quite suitable for higher education, now have the easy opportunity to get access to higher education [2][3]. Second, the limited resource into higher education because of the increasing number of students. Third, the potential changes in the composition of the students' social status, especially students with low socioeconomic backgrounds may have to work while in college to finance their studies, giving impact on academic performance [4][5].

Encouraging timely graduation is important, not least because of three factors. First, to achieve the minimum period of completion, which is one potential indicator of performance for departments, as always evaluated in a yearly quality audit. Second, the longer period of study will lead to more costs. Third, with regard to the university management and organization, for example, the lecturer-student ratios, availability of lecture room and so on. In our study case, a big private university in Indonesia experienced problems in the form of a large number of students who will lead lecturer-student ratio to be small. To maintain balance of the lecturer-student ratios, a variety of strategies are needed to cope with the perpetual students. Perpetual students problem is a common problem in Indonesia. Not only in Indonesia, it is also common in Greece, Finland and South Africa [6].

In general, the university has a policy to limit the maximum study period of 14 semesters. But in practice there are still many departments in the University is dilatory in checking the maximum period of study. As example, Department of Informatics, in 2014, have checked the maximum period of study and found that approximately 203 students has passed the 14 semesters of study. This number has been weighing on teacher-student ratios, while in fact most of these students are not active.

Enforcement of a maximum 14 semesters of study period is basically a policy pursued for the good of the institution and the students themselves. Students who graduate on time will be able to immediately find a job and settle down, moving forward in his life. Graduation for Indonesian students is not just for the pride of the students themselves, but also for families, especially the parents. While for higher education institutions, encouraging students to graduate on time will also affect the number of students, teacher-student ratios and better quality of graduates.

In this paper, we want to explain our observation study in the integration of Information and Communication Technology (ICT) to help perpetual students to graduate soon. ICT is utilized to help perpetual students because of several reasons. First, IT has a wide reach range. As the perpetual students are usually inactive students, the department has a difficulty on communicating and monitoring them. Second, ICT is able to provide both individual and mass communication. Some information for perpetual students need to be delivered in the form of mass communication, but to monitor them, the department needs more individual. Third, IT can deliver information and monitoring function in a shorter time. This short time is substantial as the department policy gives only one semester extension for perpetual students to graduate. Utilization of ICT includes control functions to monitor the progress of these students, including identifying their problems and help them find a solution of their problem.

2. Methodology

Subject of this research is Informatics department of a big Indonesian private university located in the province of Daerah Istimewa Yogyakarta. Informatics department and other departments within one faculty have big student bodies, as they didn't execute the evaluation procedure regularly. Later we found that new management of Informatics department decided to handle the perpetual students at first months of their duty as the department would be re-accredited within one year. At September 2014, the department had 38 lecturers and student body of 1869, which consist of 1697 active students and 172 non-active ones. This condition caused lecturer: students ratio as bad as 1: 49.18.

This observation study was conducted during odd semester of academic year of 2014/2015. We add one preparation month before existing semester and add one finishing month after and then break it into three periods. We call it beginning, middle and last period. Later, we divide ICT into two broad categories, one-way and two-way communications without make further details.

During August 2014 until April 2015, we recorded policies carried out by the new management and observed how they inform it to public using technology. We also observed daily technology used by the selected team to communicate with students or coordinate internally. We recorded big events such as offline gathering of perpetual students and meeting with alumni association and follow-ups using ICT. We also kept the sample of students' profile data, including academic performance, problem being discussed and effort taken by the department for each single student. We also captured some big problems and technology used by department to solve it.

Finally, we recorded every single use of each ICT per period mentioned above. Users and ICT types were the result of the research, without being specified before. At the end, we count the success rate of the department's effort.

3. Research Context

The departments' curriculum requires students to finish minimum 144 credits units to be bachelor degree, which consists of 112 compulsory credit units, minimum 21 elective credit units, 2 credit units of Community Services (mandatory), 3 credit units of mandatory Internship and 6 credit units of Final Project. There are no alternatives to finish the study except perform the curriculum. As suggested in the department academic guide, students could finish all required subjects within four years study.

The university gives maximum seven years study (excluding leaves) to get bachelor degree. It also has a policy to overcome late-study problem and prevent perpetual students by enforcing some evaluation procedures, such as Dropout (DO) evaluation at first four semesters to cut students whose GPA under threshold (2.00) and DO re-evaluation at seventh years of study.

Perpetual students of the department reached 203 students or 10.86% of total student body, and its distribution varies, as shown in Figure 1. The one and oldest student had studied for 15 years, while 67 students had seven years of study. They faced their individual problems, either academic or non-academic. Three major problems sampled from 98 persons are: 30.61% of students didn't finished compulsory credits, 28.57% didn't completed Internship project, and 7.14% of them didn't enrol Community Service yet. Non-academic problems vary from ignorance, working activities, low motivation, financial support, and family problem.

Great population, mixed grades and many individual problems enforce the department not to create a mass problem solving. To overcome this case, the department launched rescue project and decided to give them only one semester dispensation to complete their entire academic tasks. The management also allowed students to take as many as credits needed (until reach maximum 24 credit units/semester) although their GPAs were under credit passing grade. To ensure optimum result, the department assign a special task force team consisting of the secretary of department, two lecturers and two staffs.

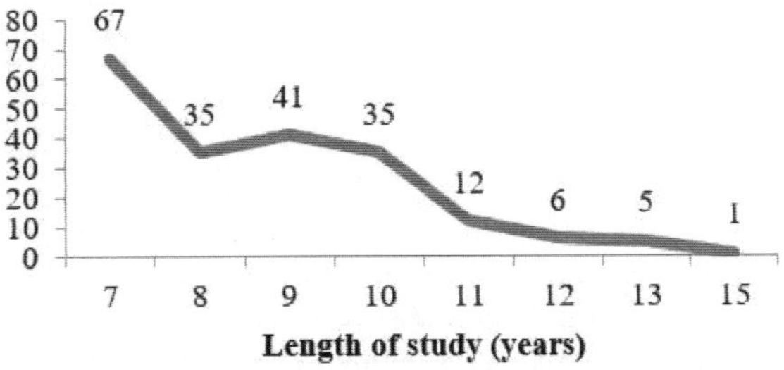

Figure 1 Number of perpetual students according to years of study

4. The use of ICT

Having a first and urgent project, the task force didn't have a master plan for Information and Communication Technology (ICT) use. But they rely on ICT, as it was hard to use paper-based solution to manage their data and to find or communicate with 203 perpetual students. Initially they used computerized academic system and spreadsheet software to check and prepare airmail to warn parents of the late study problem. By the time, types of ICT were altered and improved as the project flows. To capture more details, here we describe types of ICT use in the beginning, middle and in the end of the project, which was taken during nine months, starting from August 2014 until April 2015.

4.1. ICT use at the beginning (August – October 2014)

Students' awareness became the first challenge in this period. At first, the task force had taken the perpetual students data from online academic system supplied by the university. As it has only "flat information" which display list of student's profile (covering student number, name, GPA, total credits taken, active/non-active status, and parent information) per year admission, the team use spreadsheet software to combine and manage all information needed to process warning airmails. After analysing the data deeper, the team recognized that 37 students (equivalent with 18.84% of total) were probably impossible to be graduates as their GPA below 2.00.

The letter then typed in word processor software using mail merge features to produce many letters in a shorter time. As required by the University, all departments have to send Dropout warning letter twice before Dropout status. First letters were sent at 24 September 2014 containing several important things, such as actual student's length of study, campus regulation, length of dispensation period, and step-by-step procedure to finish study.

As many as 30 letters were returned back to campus, then the team asked other students to find unreachable students. As half of them could be contacted and then arrived to campus to update home address data, team started to recognize a kind of "friendship network" which is absolutely needed in next periods. They also found this network run on social media, using Blackberry Messenger (BBM) and Facebook.

The team also posted warning news on the department's official website which includes students' names. The news shortly became popular and reached 583 hits. But later, complaints of privacy came up and the team deactivated it a few months later.

To reach maximum perpetual students, the team also broadcasted first warning in the Facebook group of Informatics students and graduates. Perpetual students are rarely come to campus and difficult to contact with, so the department also tried to ask help from alumni association by making a special meeting and giving follow up via email containing students names. The alumni and team stay in contact using Whatsapp social media.

These widespread warnings followed by gathering invitation to build a sustainable communication and inform the department's new policy. These efforts brought good result: 48.28% of targeted students showed up in campus within a month, and most of them then came to the first gathering session at 24 September 2014.

4.2. ICT use in the middle period (November 2014 – January 2015)

Communication channels were already open, and the team decided to maintain data and interaction using ICT. Many perpetual students didn't aware campus procedures, for example they forget what password to enter online academic system, how to check unpaid tuition fee and how to register for final exam. Some of them also had low confidence and motivation to fight for graduation day. The team used online file-sharing as a coordination tool and to share data collections, which consist of students' problem and proposed solution, recent activities, city they live in, motivation to graduate, team member in charge and others.

Public channel such as website and Facebook group were not suitable in this period, so the team positioned them only as a trigger for "friendship network". This network involves offline talk (personal or group) and online through social media (person-to-person communication and discussion group in Facebook, Whatsapp and Line). Later we found that students voluntary initiated this kind of network because of social reason: solidarity and care. The team believed that better communication should give better results. Then, they concluded that private online channel such as phone call, email, SMS, and social media are greatly used to find other students and to stay contact with them. This kind of channel also used to maintain interaction between parents and the department/staff, between student/parents and final project supervisor, and also among students.

More detailed problem needs more attention, more individual and more time to solve. Online communication traffic were heavily loaded by consultation of choosing between move on or leave campus, discussion of how to set up study plan or make meeting appointment, and information request for campus' procedure (such as how to reactivate the status and fulfil administration needed for final exam). The team members also maintain offline communication. Both team and students could explain the detail of campus procedures and problem/solution that difficult to express through gadget.

In this period, the department also send second Dropout warning via airmails and publish reminder posters routinely at the Facebook group as shown in Table 1. The number of likes didn't represent the total views, but since next posts reached greater "thumbs" than previous ones, we concluded that groups members (students and graduates) became more aware of the length of study and were expected to deliver further supports for their friends. Second gathering (held at 30 January 2015) was successful because as many as 75.81% students present in the event.

Table 1. Content of social media publishing in Facebook group.

	Content of the poster	Likes
20 September 2014	Invitation for first gathering	8
13 October 2014	Last call to come back to campus	18
31 December 2014	Motivation to survive	29
4 February 2015	Information of final project exam	26
25 February 2015	Reminder of one month left to finish study	75
21 March 2015	Schedule of last final project exam for perpetual students	99

4.3. ICT use in the last period (February – April 2015)

The department administers final project exams every month. A number of 124 students worked on final projects during this dispensation semester. Luckily, one-third students have GPA more than 3.00; so the team assumed academic competence was not a problem. Unfortunately, although exam schedule had been announced since first gathering at September 2014, almost sixty students didn't register the exam until February 2015. As March 2015 became the last exam for odd semester, the task force team had a very difficult task to ensure candidate students to fulfil the requirement for final exam registration and register for exam.

During this period, the department delivered mass communication twice only at Facebook group. Furthermore, team members check one-by-one graduate candidates more intensively. Existed alumni and fresh graduates from perpetual students surely help the team to check-and-contact friends personally. All of them utilized both conventional (phone call, SMS) and private channel in social media (Blackberry messages, Whatsapp, Line, and Facebook). Finally, 57 perpetual students registered for the last exam period and 91.23% of them passed the exam.

Furthermore, students who failed to be graduates use email, Whatsapp, and SMS to ask procedure and process to go out or move to another campus. The team also supply the required documents in the department's official website. At the end, 53.01% perpetual students graduate successfully, 15.66% choose to leave campus, and 31.33% become dropouts as communication never exist or GPA is under the threshold. It means the team had successfully maintained communication at level 70%.

5. Result and Discussion

Ineffective students management leads to negative effect for the ratio of lecturer: students and other problems. The use of ICT for rescue project in the research could be summarized in the Table 2.

Table 2. Various use of ICT to support perpetual students

		Participants	1st period	2nd period	3rd period
One-way					
	Department website	Sender: the team Recipient: students, graduates, parents	To announce warning	To announce warning, list participants of final project exam	To announce warning, list participants of final project exam, share document for leaving campus
	Academic information system	User: team members	To get students academic profile	To inform recent student's academic status	To inform recent student's academic status
	Desktop-based spreadsheet	User: team members	To record data taken from academic IS	-	-
	Word processing	User: team members	To print first warning letters	To print second warning letters	-
Two-way					
	Online file-sharing	User: team members	-	Data sharing and coordination tool	Data sharing and coordination tool
	Airmail	Sender: the team Recipient: parents	To notify parents	To notify parents	-
	Phone call/SMS	Teams, students, graduates, parents, supervisor	Personal consultation and discussion	Personal consultation and discussion	Personal consultation and discussion
	Email	Sender: the team Recipient: alumni association	To ask help to find students	-	To share document for leaving campus
	Facebook group	Sender: the team Recipient: students, graduates	To post info/motivation/reminder/warning/	To post info/motivation/reminder/warning	To post info/motivation/reminder/warning
	Social media: Official Facebook/Whatsapp members owned by department/team	Teams, students	-	Personal consultation and discussion	Personal consultation and discussion
	Social media: BBM/Facebook/Whatsapp/ Line	Among students, between students and team/supervisor	To find, inform, remind, and motivate students	To find, inform, remind, and motivate students	To check exam preparation, inform, remind, and motivate students

ICT comes in various types and functionalities. Computerized academic system and department's official website are common example of formal and one-directional media to take the authorized data from and to broadcast information to public (perpetual students, alumni, and parents). To deliver accurate information for parents, sending airmail letters are also necessity. Creating and formatting those formal letters become more efficient while using office software such as word processor and spreadsheet.

One-directional communication channel such as department's website and email to alumni is a good starting point to contact perpetual students. But we must also keep them in contact using private channel, especially social media to ensure they run in the right track and finish the study as expected. As perpetual students usually face academic and non-academic problems, good communications among stakeholder become well deliverables using private channel. Topics vary from asking a very simple procedure to counselling study plan in a short time. But remember, as private channel become dominant in the middle and last period, assigned lecture/staff team should be more knowledgeable, patient and robust in these periods.

Learning from publishing poster in Facebook group, we conclude that being a trustworthy and reputable department for students or alumni need process and time. It is also necessary to collaborate broadcast and private channel periodically.

Furthermore, we recommend department to use academic information system intensively and assign special team to handle late study problem, which equipped by various ICT means. As every person is unique and probably using different technology, we also recommend to create such a master plan for each ICT channel to get higher success rate.

References

[1] Katsikas A, Panagiotidis T 2011 Student status and academic performance: Accounting for the symptom of long duration of studies in Greece *Studies in Educational Evaluation 37 (2011) 152–161*

[2] Shofer E, Meyer J 2005 The worldwide expansion of higher education in the twentieth century *American Sociological Review* 70(6) 898–920

[3] Trow M 2007 Reflections on the transition from elite to mass to universal access: Forms and phases of higher education in modern societies since WWII *Forest, J., & Altbach, P. (Eds.), International handbook of higher education* Vol. 18 (Springer) pp. 243– 280.

[4] Bratti M, Checchi D, de Blassio G 2008 Does the expansion of higher education increase the equality of educational opportunities? Evidence from Italy *IZA Discussion Papers* 3361, Institute for the Study of Labour

[5] Metcalf H 2003 Increasing inequality in higher education: The role of term-time working *Oxford Review of Education* 29(3) 315–329.

[6] Kalamatianou A G, McClean S 2003 The Perpetual Student: Modeling Duration of Undergraduate Studies Based on Lifetime-Type Educational Data *Lifetime Data Analysis* 9, 311–330.

Using Isovist Application to Explore Visibility Area of Hospital Inpatient Ward

M M C Sengke[1] and P Atmodiwirjo[2]

Department of Architecture, Faculty of Engineering, Universitas Indonesia, Kampus Baru UI, Depok 16424, Indonesia

[1]maria.myron@ui.ac.id and [2]paramita@eng.ui.ac.id

Abstract. This paper reports an on-going project that explores the use of digital application to study human field of view. The focus of discussion is to study the patients' visual experience, in relation to the arrangement of interior elements in patients' ward. The physical qualities of the environment can influence the healing process of the patient. Typical layout of interior elements often fails to provide visual stimulus that could support the healing process of the. This study explores the experience of seeing by simulating the hospital ward setting into 3D model using isovist analysis. Isovist is used to represent the experience of seeing by the patient from particular point of view and also to represent the object and surfaces that are being seen. Isovist has a function to show us the boundary of the visible areas, which can reveal which elements can and cannot seen by the patient. Isovist provides a way to understand the experience of seeing and being seen by visualizing the visibility area through three dimensional modelling. This study suggests the possibility to study human field of view to support the design of architecture for health.

1. Introduction

The purpose of this study is to explore visibility area of hospital inpatient ward through isovist application. Visibility area becomes important in healthcare architecture, since many research revealed the effect of viewing to patients health condition. Ulrich emphasized that seeing experience may influence patients recovery [1]. He discussed the limitation in hospital to give visual access for the patients [1]. Access to seeing experience is a way to provide visual information which could stimulate attention from the patient

In general the design guidance for healthcare buildings already suggested the importance of visual access for patient. This includes the views out of the building, design to give people contact with nature, finishes that provide a variety of colours and texture, art to provide stimulation and distraction, and interior appereance design to inspire confidence and positive atmosphere [2]. However there is a question of how the patients experience the activity of seeing, and how the environment of the spatient plays a role in providing visual information from the arrangement of interior elements.

There are many methods to analysis the visual information in an environment such as isovist approach. Isovist has been used for various purposes of application in various fields including in studying the perception of space which is in the intersection between the fields of psychology and architecture. Benedikt explained the relevance of isovist to architecture and isovist "as a collection of visible area real surface in space" [3]. He also described an isovist as "the set of all points visible form a given vantage point in space" [3]. Turner denoted isovist as "a viewshed or an area in spatial environment directly visible from a location within the space" [4]. Batty explained isovist as "the space that can be seen from any vantage point" [5].

Some research have used isovist as a visual analytical tool to explore the visibility area. For example; to study the pattern of connections for a simple spatial configuration [4]. and to study the visibility degree of spatial configuration [6]. This research attempts to examine the negative and positive space experienced visually through isovist 3D model, which indicates the seeing experience created by the arrangement of interior elements in the patients' ward in the hospital.

Isovist approach could reveal that human field of view can be represented in particular shapes. The shapes of human field of view indicate what is detectable by the viewpoint in a visual medium from one point and represent their geometric properties. Isovist can be used to visualize the visibility area represented through a graph of visibility area in geometric features. Therefore, it needs the data of spatial and human field of view to determine the boundary of the visibility area.

Human field of view can be known by exploring vision experience through vision mechanism. Gibson defined the term field of view as a viewpoint that occurs from the reflection of light which is visible to the eye [7]. The field of view has a boundary that defines certain large visual area. The angular area of the field of view depends on the position of eyes, head and body in the medium of vision. The essential elements of human field of view are its boundaries, both obscure and boundless boundaries. These boundaries differentiated the visible area from the invisible area. Essentially, the field of view indicates whatever the visual world presents to the observer and detectable to the eyes through the field of view

2. The Role of Human Field of View in Experiencing Space

The experience of seeing and being seen is intertwined and play a role as a part of psychoneuroimmunology, in which psychic, mind and body had interplay with the environment as a part in the whole of human wellbeing [8]. Trevelyan explain "seeing as experiencing" as a spatial experience through visual experience [9]. Eye movement traces elements of space, from the plane of the wall, to a pillar and then to the column and into the wall above the opening and then to the openings Visual experience brings the patient from one image to another image that together bring the views of life.

"Seeing as healing" occurs when the experience of seeing can influence recovery process through the response of the patients' sensory that have been triggered or controlled by the spatial elements in the environment [10]. Investigation of the patients' responses to visual environment such as views of natural scene, art products, and plasma windows has shown a strong effect such as reducing fear and stress, replacing with positive distraction, holding interest, evoking positive feelings and curiousity [11]. A study by Ulrich found that patients in pain and suffering condition in patient's room with natural window tend to heal in shorter time, have fewer complaints and have lower level of minor complication after surgery. In addition, he also found the different conditions of patients with wall-view in the ward compared to those with the tree-view [12].

This paper explores both the mechanism of seeing and being seen as experienced by patients in healthcare environment. In the mechanism of seeing, human field of view involved the eye structures such as focus function view (fovea vision) and the spread of the focal point of the view (peripheral vision) that create the visual range (field of view) [13][14][15]. Gibson explained that human field of view is influenced by the position of the observer, which shows the existence of the self and the motion of the body [7]. In the mechanism of seeing, isovist is used to analyze the visual angle which includes the position of human field of view.

Environment as being seen takes a role as stimulus by providing information through the reflection of light on the surface that humans can respond. In this process, the transactions happens through information provision and information reception. The surface could be visible if there is a light source and if it has the ability to reflect light [7]. In the mechanism of being seen, isovist is used to record the surface through the advantage point in the medium of vision.

3. Methods

This study used a single isovist method to explore the mechanism of seeing and being seen. The method was applied by using three-dimensional modeling program Rhinoceros which is supported by the graphics algorithm of Grasshopper. Digital modeling was used to represent the existing condition of the patient's room in reality. The isovist approach was applied to reveal which elements can be seen by the patient in a lying position and also to find the characters of the boundary of visibility area model.

The study was performed on a set of virtual indoor scenes of hospital ward setting as the stimuli of human visual of field. The scenes represented spatial situations in an inpatient ward which contains four patient beds, cabinet, curtain, and some medical equipment. This study applied two methods to represent the stimuli for human field of view. The first is by using the camera viewpoint in 3d model created in Rhinoceros. Angles of view indicate different directions of the neck, head and eye movement. Every sections are modelled based on the direction of y or z-axis. The direction of y-axis shows the motion of viewpoint from buttom to the top view. The viewpoint based on the z-axis shows the motion of viewpoint from left to the right view. By using this method it is possible to register a series of visual scenes as a simulation of the way patients view their environment.

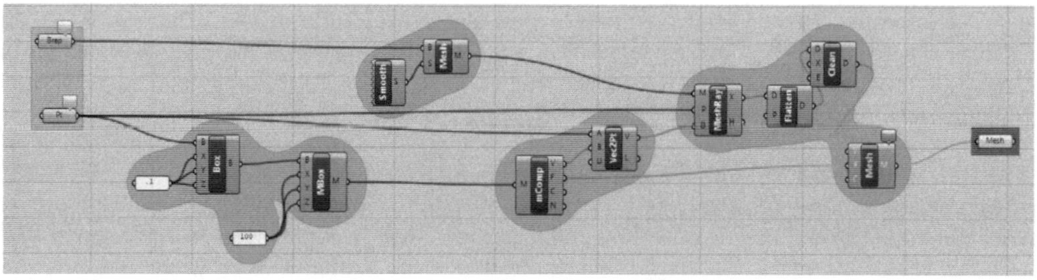

Figure 1. 3D Isovist formula in Grasshopper [16].

The second method is by using isovist application in three dimensional modelling with Rhinoceros with Grasshopper plug-in. The 3D isovist formula in Figure 1 which is used in this method was adapted from Heumann and then applied in the model in Figure 2 which is created in 3D model Rhinoceros that represents the actual condition of inpatient ward in hospital [16].

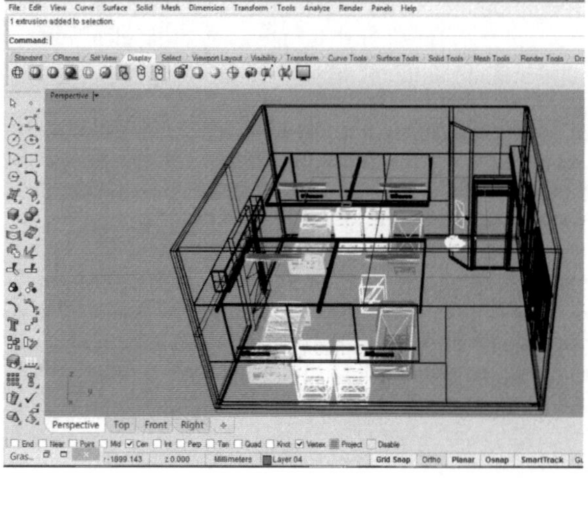

Figure 2. 3D Modelling in Rhinoceros, Sengke and Atmodiwirjo (2016).

In applying the 3D isovist formula, Boundary Representations (Brep) was set to the 3D model and Point was set to the Point in 3D model so it will make a link and create the 3D isovist in the 3D model. A point was set on the position where the eyes are located. In this case the position of the eyes were adjusted with the patients position while lying on the bed. By using isovist analysis in one point of view, it is possible to detect the visible area and generate the shape of that area.

4. Findings

The findings indicates the visibility area of the patients into two parts: the analysis of seeing from certain point of view and the analysis of space that is being seen.

4.1. Analysis of Seeing from Certain Point of View

This analysis revealed the elements interior that are within the patient's field of view when seeing happen in a movement along one axis. Coinciding some interior elements did not found in the range of the patient's visual. That is mean elements interior are in the area of the blind spot. The elements such as: floor surface, the surface area of the wall behind the patient, the head of the bed, the identity on the front bed and hospital equipment that are placed beyond the visual range. This condition may caused the patient is lack of information.

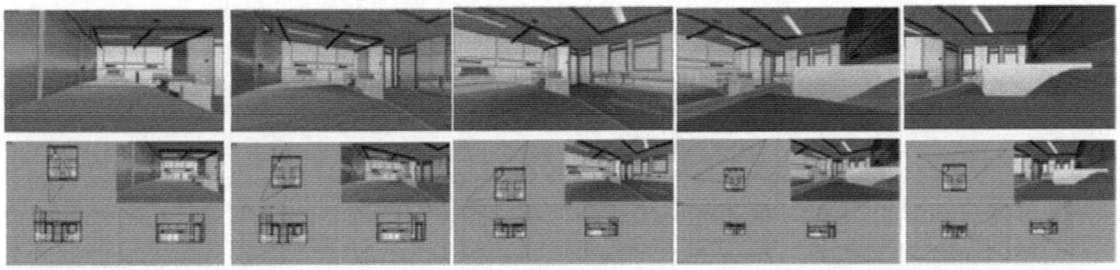

Figure 3. Perspective Scene in the z-axis motion and the motion camera angles for the z-axis produces the movement from left to the right view, Sengke and Atmodiwirjo (2016).

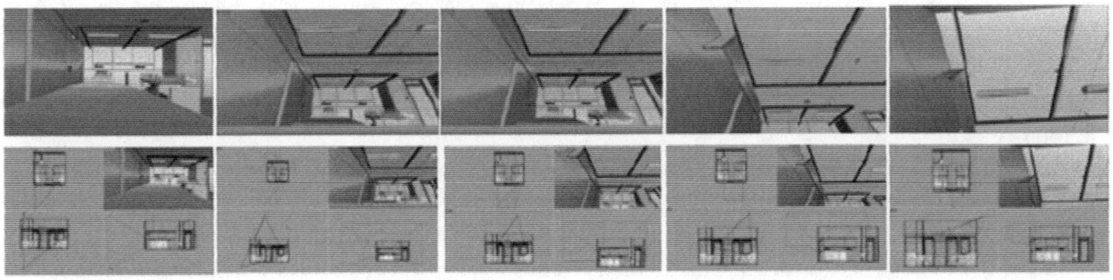

Figure 4. Display images in the visual range based on on the y-axis motion which produces the movement from the front to the top view and the motion camera angle with y-axis, Sengke and Atmodiwirjo (2016).

4.2. Analysis of Space that is Being Seen

The second analysis attempts to simulate the geometric shapes of the patients field of view. The boundary of visibility area shows which area can be seen and which can not be seen by the patient in a lying position on a bed in the inpatient ward. This boundary area show the visual capacity provided by the environment of the patient. Viewpoint from one advantage point in isovist application create the medium of visibility area.

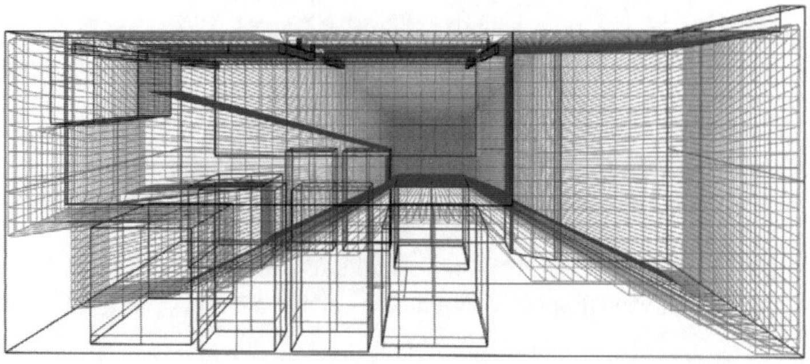

Figure 5a. The point of view that show a wider field of view than the other, Sengke and Atmodiwirjo (2016)

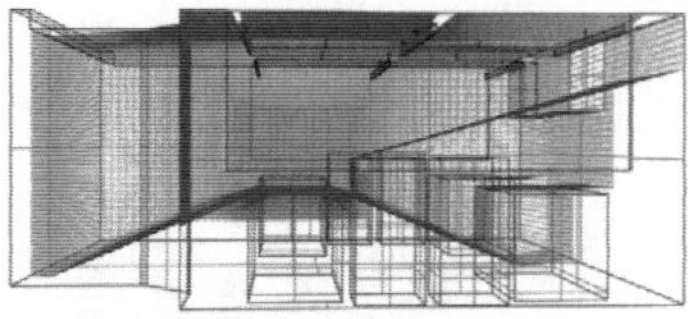

Figure 5b. The Position of the bed that has a field of view close to symmetry, Sengke and Atmodiwirjo (2016)

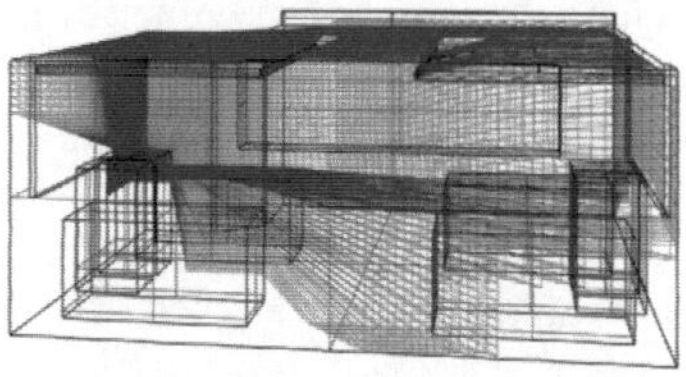

Figure 5c. The Characteristics of interior elements limit the field of view of patient, Sengke and Atmodiwirjo (2016)

This case shows that some elements were found to be the barriers that limit the field of view of the patients. Some characteristics of the elements such as the height of the cabinet could block the view. The placement of the patients' bed causes some bed to have a wider view than the others. Through isovist analysis, it is revealed how the layout of interior elements in inpatient ward could affect the human visual of field from each patient's position. The interior elements arrangement in inpatient ward, could create a different view for each patient, which may affect their visual experience of space and affect their wellbeing.

5. Conclusion

This on-going research has illustrated that the isovist application could represent the experience of seeing by the patient from particular point of view and also represent the boundary of the visible areas, which can reveal which elements can and cannot seen by the patient. Isovist provides a way to understand the experience of seeing and being seen by visualizing the visibility area through three dimensional modelling.

This study has presented an isovist analytical approach based on the capability of seeing experienced by the patient in bed-ridden condition. The result shows the visibility area of patient which includes the visibility surface and visibility volume from the patient environment. This visibility area plays a role to provide a series of visual information for the patient. The findings of the study shows the interior elements included in the patient's field of view and indicates the surface area that are visible. By understanding the visible surface and interior elements, it is possible to define the positioning of visual information which could stimulate attention from the patient, improve the experiencing of seeing, and give a positive distraction to the patient.

The use of isovist application in studying healthcare environment could enhance the relationship between architecture and health and humans and their environment. Further study will develop this analysis by looking in more details the field of view of patients in hopital ward in relation to the presence of architecture and interior objects or elements.

6. Acknowledgments
This study was funded by Universitas Indonesia Research Cluster Grant 2015 for Hospital and Healthcare Design and Engineering Research Cluster.

References
[1] Ulrich 1984 *View Through a Window may Influence Recovery From Surgery* (*Science* Vol 224) p 420
[2] Department of Health 2014 *Health Building Note 00-01 General Design Guidance for Healthcare Buildings* London
[3] Benedikt M L 1979 *To Take Hold of Space Isovist and Isovist Fields* (*Environment and Planning B* Vol 6) p 47-65
[4] Turner A Doxa M O Sullivan D & Penn A 2001 *From Isovists to Visibility Graphs a Methodology for the Analysis of Architectural Space* (*Environment and Planning B* Vol 28) p 103-121
[5] Batty M 2001 *Exploring Isovist Fields Space and Shape in Architectural and Urban Morphology* (*Environment and Planning B* Vol 28) P 123-150
[6] Johanes M Yatmo Y A & Atmodiwirjo P 2015 *The Use of Computational Medium for Visualization and Simulation in Healthcare Architectural Design* (Research Cluster Grant Indonesia)
[7] Gibson J J 2014 *The Ecological Approach to Visual Perception* (New York and London: Psychology Press)
[8] Suresh Mini and Smith Dianne and Franz Jill 2006 *Person Environment Relationships to Health and Wellbeing An Integrated Approach* (IDEA) p 87-102
[9] Trevelyan S G 1977 *The Active Eye in Architecture an Approach to Dynamic and Imaginative Seeing* (Wrekin Tust: Ross on Wye-Herefordshire)
[10] Nanda U Chanaud C Nerlson M Zhu X Bajema R Jansen B H 2012 *Impact of Visual Art on Patient Behaviour in the Emergency Department Waiting Room* (*The Journal of Emergency Medicine* Vol 43) P 172-181
[11] Ho T H Rainbow Potash S Jordan Fang Fan (2015) *Art Viewing Directives in Hospital Setting Effect on Mood* (HERD Vol 8(3)) p 30-43
[12] Ulrich R S 1993 *Biophilia Biophobia and Natural Landscapes* (Washington D C: Island Press) p 73-137
[13] Pheasant S 2003 *Bodyspace Anthropometry Ergonomics and the Design of Work* (USA: Taylor & Francis)
[14] Yarbus A L 1967 *Eye movements and Vision* (New York: Plenum Press)
[15] Yanthis Steven 2014 *Sensation and Perception* (USA: Worth Publisher)
[16] Heumann A 2010 *3d Isovist* http://heumanndesigntech.wordpress.com

Development of Multi-slice Analytical Tool to Support BIM-based Design Process

P Atmodiwirjo[1], M Johanes[1], Y A Yatmo[1]

[1] Department of Architecture, Faculty of Engineering, Universitas Indonesia

E-mail: paramita@eng.ui.ac.id, mikhael.johanes@eng.ui.ac.id, yandiay@eng.ui.ac.id

Abstract. This paper describes the on-going development of computational tool to analyse architecture and interior space based on multi-slice representation approach that is integrated with Building Information Modelling (BIM). Architecture and interior space is experienced as a dynamic entity, which have the spatial properties that might be variable from one part of space to another, therefore the representation of space through standard architectural drawings is sometimes not sufficient. The representation of space as a series of slices with certain properties in each slice becomes important, so that the different characteristics in each part of space could inform the design process. The analytical tool is developed for use as a stand-alone application that utilises the data exported from generic BIM modelling tool. The tool would be useful to assist design development process that applies BIM, particularly for the design of architecture and interior spaces that are experienced as continuous spaces. The tool allows the identification of how the spatial properties change dynamically throughout the space and allows the prediction of the potential design problems. Integrating the multi-slice analytical tool in BIM-based design process thereby could assist the architects to generate better design and to avoid unnecessary costs that are often caused by failure to identify problems during design development stages.

1. Introduction: Multi-slice representation in architecture

Architecture design process aims to represent the ideas of space to be materialized into built environment. Architecture is experienced by human being as a dynamic entity; it involves movement from one space to another, and it involves changes of spatial properties from one time to another [1]. One of the challenges in representing the architectural ideas is on how to depict architectural space as a dynamic entity, which have the spatial properties that might be variable from one part of space to another. The standard architectural drawings, which are primarily based on the projection of space at certain point and at certain time, are often insufficient to depict the dynamics of the space.

This paper attempts to extend the techniques of architectural representation through multi-slice representation approach. Multi-slice representation refers to the technique of tomography as practiced in medical science, in which the three dimensional human body is analysed by making a series of slices that together provide a comprehensive image of the human body [2]. The understanding of space as a series of slice is important in architecture, especially to analyse the types of space that involves dynamic movement, in which the change of spatial properties and the continuity of space could affect the performance of the space. As an attempt to apply multi-slice representation approach in the design process, a multi-slice analytical tool is developed and integrated into Building Information Modelling.

Building Information Modelling (BIM) approach transforms the documentation method in architectural production from two dimensional drawing information systems into three dimensional object information systems [3]. By using more integrated system, BIM approach could avoid the fragmentation and the duplication of data which could lead to confusion and inefficient workflow. As a result, BIM adoption in architectural and construction industry becomes important to increase productivity, efficiency, quality and sustainable development [4]. Motivated by such immense development of BIM approach in architectural design, we attempted to develop multi-slice analytical tool that is integrated with BIM environment in order to support architectural design process.

The purpose of this paper is to give a brief description of the development of BIM analytical tool that could represent architecture and interior space as a series of slices based on particular paths that are experienced along the space. This paper takes an example of the use of multi-slice analytical tool to analyse a hospital corridor space, in which the continuity of performance along the space is important. Through multi-slice representation, it is possible to gather detailed spatial information in each slice, which will be important to assist the design process in which the spatial continuity becomes the main consideration.

2. The development of multi-slice analytical tool

Rather than developing the analytical tool specifically for particular BIM-based modelling program [5], we used a more flexible approach by creating a stand-alone application to extend its functionality (Figure 1). The analytical tool uses Industry Foundation Classes (IFC) data that is exported from the Autodesk Revit modelling software. IFC is an open neutral format to describe industrial data from building and construction, developed by BuildingSMART based on ISO-STEP EXPRESS language and concepts [6]. Although improvement is still needed to reach fully effective interoperability, IFC currently becomes the state-of-the-art exchange file format among different BIM platforms [3]. IFC is developed as collections of various entities (data objects), property sets and data types that represent the geometrical and semantic richness of building information. IFC was designed to address all building information that could be used for different purposes, such as energy analysis, cost estimation and scheduling [6]. Such interoperability in BIM approach expands the possibility of BIM-based tool development that could address more specific tasks that could not be provided by generic BIM tools.

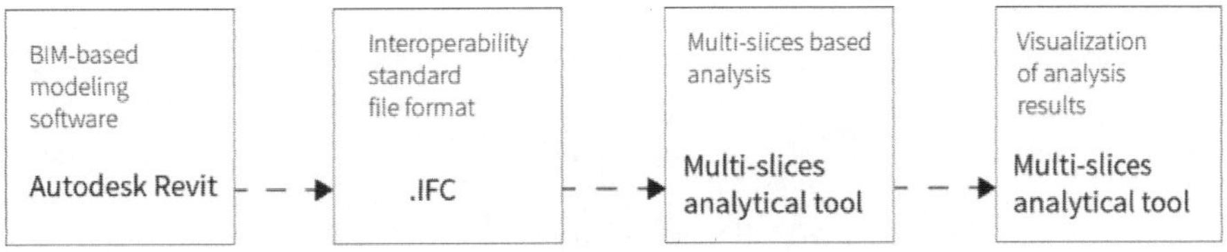

Figure 1. The general workflow of multi-slice analytical tool

The development of multi-slice analytical tool was done using Python scripting language due to its vast libraries and community. Most parts of the tool are built based on IfcOpenShell combined with PythonOCC to process the building information contained in IFC files. IfcOpenShell is an open source library that uses 3D CAD Kernel OCE (Open Cascade Community Edition) to convert the geometry information from IFC file into explicit geometry that can be understood by another CAD or modeling library. The resulted geometry and its relating information such as materials and its semantic are then processed and analyzed using PythonOCC robust modeling algorithm. PythonOCC itself is a 3D CAD development framework that also builds upon OCE.

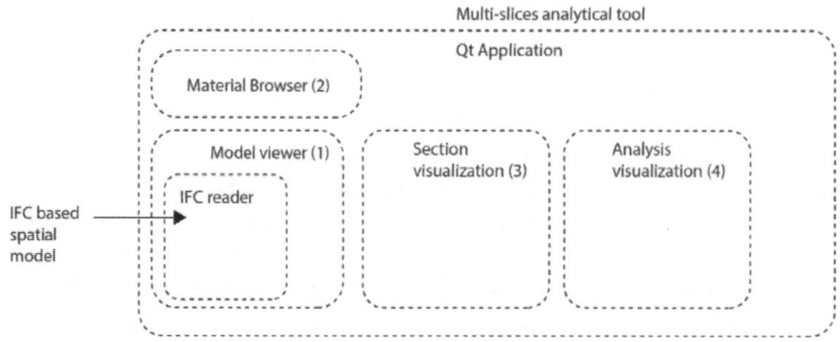

Figure 2. Multi-slice analytical tool modules

The analytical tool is developed as a stand-alone package that uses the spatial information from the IFC files provided by other modeling programs (Figure 2). The tool consists of three main parts; The Model Viewer (1), Section Visualization (3) and Analysis Visualization (4). Model Viewer reads the IFC file into memory and displays it on interactive three-dimensional display. In this part the user needs to determine which part of the model that would be analyzed by generating the series of sections. The Section Visualization module rearranges the generated sections into a parallel sequence. The tool thereby could perform some analyses from the series of section to assess its spatial continuity based on some properties such as spatial dimension, materials, position of elements, etc. The analytical tool visualizes the results of analyses in two-dimensional representation in Analysis Visualization module to provide better readings from the results of analyses.

3. The use of multi-slice analytical tool

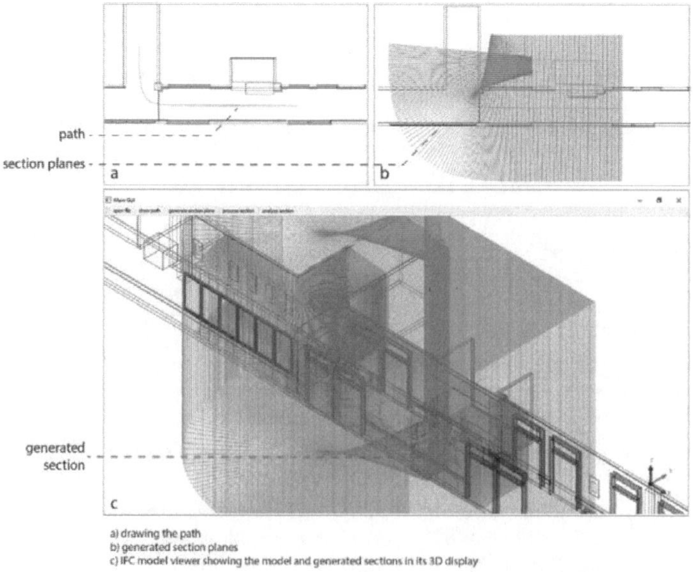

Figure 3. Series of slices generated on IFC model viewer screen

The following example demonstrates the use of multi-slice analytical tool to analyze the space of hospital corridor. Figure 3 shows how the slices are generated on Model Viewer. The analysis is performed by drawing the path as the main axis to generate the series of slices. The sections are generated perpendicular to the path drawn by the user on the viewer. The following pseudo-code shows the process of generating the sections based on resulted section planes and building elements on IFC model.

SECTION-SLICES(*buildingElementList, sectionPlaneList*)
declare *sectionList*
for each *sectionPlane* in *sectionPlaneList*
    sectionElements ← CREATE-SECTION(*buildingElementList, sectionPlane*)
    section = [*sectionElements, sectionPlane*]
    insert *section* **to** *sectionList*
return *sectionList*

CREATE-SECTION(*buildingElementList, sectionPlane*)
declare *sectionElementList*
for each *buildingElement* in *buildingElementList*
    sectionGeom ← GET-INTERSECTION-CURVES(*buildingElement, sectionPlane*)
    sectionElement ← [*sectionGeom, buildingElement*]
    insert *sectionElement* **to** *sectionElementList*
return *sectionElementList*

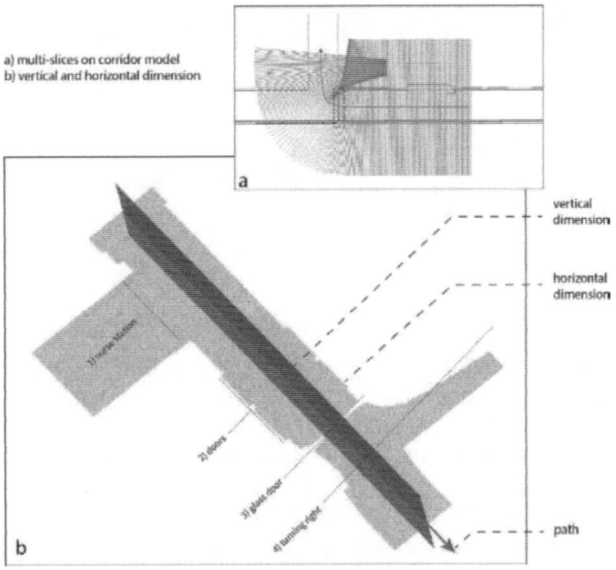

Figure 4. Parallel sequence of slices on Analysis visualization screen

The generated series of sections are then translated into parallel sequence to allow the spatial comparison between and within the sections [7] from the beginning to the end of the path (Figure 4). The red lines and blue lines are created from the center of section plane to measure the boundary of the space along its axis. The measurement is performed by finding the nearest intersection from origin point of the section plane to section curves along each axis. The resulted axis lines provide general measurement of corridor dimension but not showing the clearance dimension of corridor space.

Figure 5 shows the principle of calculating the space clearance on each $section_j$. The measurement of the horizontal clearance is performed by finding the shortest distance between the center and both left and right sides of each section from bottom k_{min} to the top k_{max} of the boundary. The *horizontal* $left_{jk}$

and $\textit{horizontal right}_{jk}$ denotes the horizontal clearance on index k on $\textit{section}_j$. The vertical clearance measured by measuring the distance $\textit{vertical distance}_{ji}$ between bottom and upper side of each section from left i_{min} to right i_{max} of the boundary.

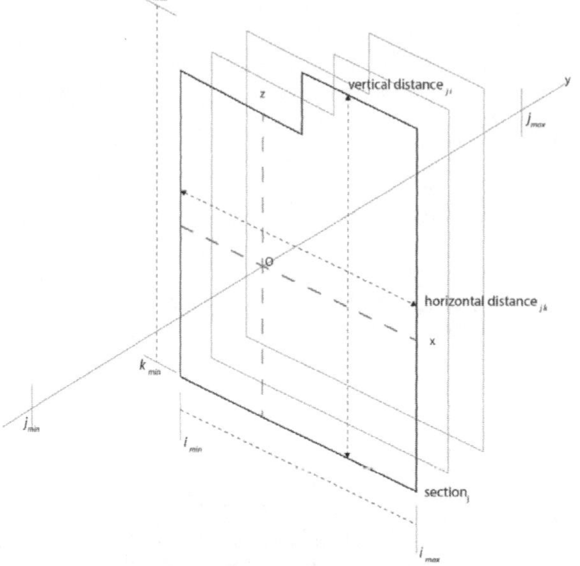

Figure 5. Measurement of space clearance on section

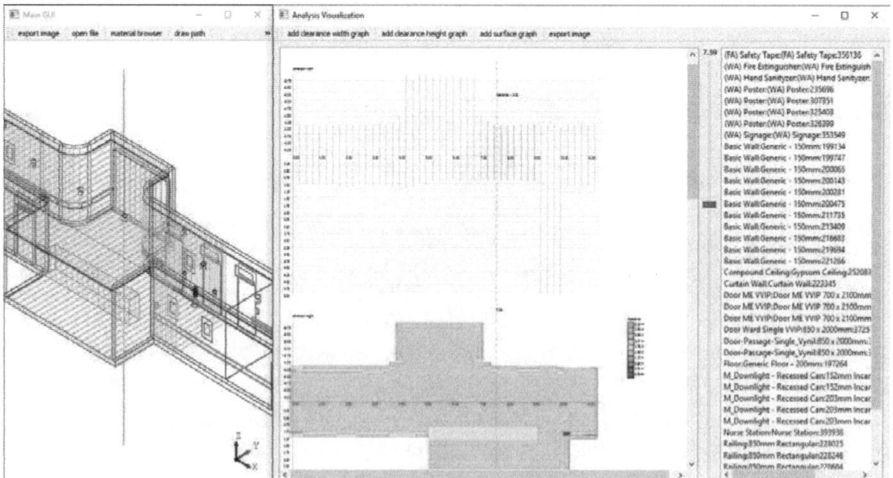

Figure 6. Visualization of space clearance analysis

Figure 6 shows the visualization of space clearance analysis of the corridor. The horizontal clearance is shown as a series of vertical lines spreading from zero axis that represents the clearance distance of each section. The green red lines indicate that the clearance is below the minimum clearance that is required by the standard, while the green lines indicate the opposite situation. The color-scaled graph represents the vertical clearance that shows the distance between the upper part and the bottom part along the whole corridor. The distance is shown by the color-scale, in which red represents the shortest measured distance, while green represents the farthest measured distance. Both graphs show the void space that allows people to move along the corridor, which is difficult to illustrate on standard architectural drawings. By using the space clearance graph, the architects could have better feedback on the performance of the analyzed space.

4. Possible further development

The development of multi-slice analytical tool in integration with BIM allows the analysis of architectural space for its continuity of performance. This development is a part of ongoing research that looks into how the analytical tool could be used to assess the performance of hospital corridor, in particular to assess the spatial elements that could affect the spatial continuity of the space, such as handrails, finish materials, position of doors, and lighting.

Some further development is necessary to improve the quality of the multi-slice analysis to provide better feedback for design process. In multi-slice analysis, the distance between sections correlates directly with the resolution of spatial information that would be analyzed. Narrowing the distance between sections will result in larger number of section slices thus providing more complete representation but increasing the computational cost. Further study could attempt to find the relationship between distance between sections and collected spatial information in order to determine the optimal number of slices needed to perform the analysis.

Further development is also necessary to improve the features of the tool in order to better support design development process. The semantic richness of IFC opens the possibilities of more detailed analysis on every spatial element in their context. While the tool development currently focuses on the spatial analysis, the visualization of the analysis is also important since the multiplicity and density of displayed information [7] might affect how the analytical tool would give insight about the analyzed space. The development of such visualization will provide a valuable data to support the design development process.

References

[1] Jones P B and Meagher M 2014 *Architecture and Movement: The Dynamic Experience of Buildings and Landscapes* (Routledge)

[2] Kelly B 2014 Medical Transplant Proceedings of Association of Collegiate Schools of Architecture 102nd *Annual Meeting: Globalizing Architecture/Flows and Disruptions* (Miami Beach, Florida) pp 615-623

[3] Jeong Y S, Eastman C M, Sacks R and Kaner I 2009 Benchmark test for BIM data exchanges of precast concrete *Automation in Construction* **18** pp 469-468

[4] Arayici Y, Coates P, Koskela L, Kagioglou M, Usher C and O'Reilly K 2011 Technology adoption in the BIM implementation for lean architectural practice *Automation in Construction* **20** pp 189-195

[5] Doherty B, Rumery D, Barnes B and Zhou B 2012 A Spatial Query & Analysis Tool for Architects *Symposium on Simulation for Architecture and Urban Design* (Florida) ed L Nikolovska and R Attar pp 31-38

[6] Eastman C, Teicholz P, Sacks R and Liston K 2011 *BIM Handbook: A Guide to Building Information Modeling for Owners, Managers, Designers, Engineers and Contractors* (John Wiley & Sons, Inc.)

[7] Tufte E R 1990 *Envisioning Information* (Connecticut: Graphic Press)

IOP Conf. Series: Materials Science and Engineering **185** (2017) 012010 doi:10.1088/1757-899X/185/1/012010

The Impact of Islamic Messages on Twitter Towards Moslem Youth

[1]Ahmad Fathan Hidayatullah, [2]Almed Hamzah

Department of Informatics, Universitas Islam Indonesia
Jalan Kaliurang KM 14,5 Sleman Yogyakarta Indonesia

Email: [1]fathan@uii.ac.id, [2]almed.hamzah@uii.ac.id

Abstract. The presence of social media has increased the opportunities for the Moslem in sharing and gaining knowledge about the religion of Islam. In general, Twitter can be an alternative for people to deliver and obtain Islamic messages. As the biggest moslem community, Indonesian people has made use of Twitter as a tool for them to gain knowledge about religion. As a matter of fact, there are many Islamic preachers that have been utilize Twitter as a media of proselytizing various topic related to religion practice. The most popular preacher usually has many follower. The result shows that there is a strong correlation between the two. There is a correlation between the number of tweets with the number of follower. The more preacher post a tweets the more follower they attained. Moreover, several topics are identified as the most frequently occurred. The tweets which contain about worship, love, information, and purification of soul have become the most favourable subject among preachers and scholars.

1. Introduction

Among the number of social networking sites used by people in Indonesia, Twitter is the one of the media that have been utilized for spreading and gaining information about Islam. Compared with other social networking sites, Twitter has special characteristics in its message. Twitter allows user to share information through 140 character of short message called tweet. However, it is very often that users write many messages in the same topic with length more than 140. Therefore, they will divide the messages into a series by using hashtag (#) followed by the topic or by numbering the tweets. Figure 1 below shows the example of Twitter message series from the account @RumayshoCom with numbers and the hashtag #rezeki which means provision from God.

In Indonesia, there is a term called kultwit which stands for "Kuliah Twitter". It is defined as a set of tweets that belong to certain topic or theme. Additionally, the purpose of kultwit is to deliver a set of message from particular topic [1]. The topic itself ranging from various area such as politics, religion, art, and so on. However, this study will focus on the messages with religion (Islam) topic. In general, Islamic messages on Twitter usually delivered by a person who has proper knowledge about Islam which is called da'i (the person who do preaching).

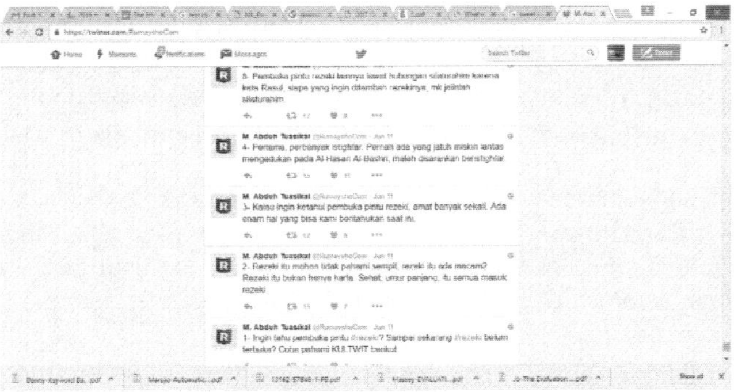

Figure 1. The example of Islamic Tweet Series

Based on the previous research [1], it is revealed that spreading Islamic messages via Twitter by using kultwit has become trends among some Islamic scholars in Indonesia. In addition, this also help them to convey their messages with more attractive to all people and specifically to the Islamic community. However, there are spaces for research to enquire this phenomena deeper.

Thus, this study will discussed about two things. Firstly, this research will find the correlation between the number of the scholar followers and the enthusiasm of the follower towards the frequency of tweets posted by the scholars. Secondly, we retrieve the Twitter messages from the most famous scholars which mentioned by our respondents in the questionnaire. Moreover, we also cluster the tweets to know the tweet content posted by the scholars.

The rest of this paper will described as follows. Second part of the paper will be discussion about related works that has been done by other researcher. Third part will be method used in this study. Fourth part will be the result of the study along with the discussion related to it. Finally, concluding remark will end this paper.

2. Related Works

Some previous research have been discussed about the utilization of social media for religion purposes.

Lovheim [2] has revealed the high number of internet usage among young people in Sweden. However, the use of internet to find information related to religion still low (40%). In addition, this study stated that the internet will become new place for the contemporary people to discuss anything about religion in the future which is called cyber religion.

Hosseini [3] has investigated about the role of information technology (website and internet) for preaching activity. The result of the study stated that information technology was able to provide support in conveying the knowledge about tajweed, hadith, tafseer and another contemporary topic. The existence of information technology somehow is able to enrich the user's understanding about Islam. This is achieved by reading literature online. The advancement of information technology should have been used to spread Islamic messages [4].

Adam, et al. [5] has studied the use of blog as a media for Islamic preaching) in Malaysia. Blog has been adopted by the scholars to share islamic knowledge. Moreover, it has also been regarded as the one of effective alternative methods in disseminating any information related to Islam. In spite of that, the use of the blog to look for information about religion is still low among young people. They perceived blog as media only for entertainment and socializing with people around the world, not to find a particular knowledge.

Agboola [6] stated that the use of website in educative preaching is still low. This is due to the fact that only few people contribute in it. Furthermore, recently, internet users are more attracted by social media rather than website. Thus, to utilize social media in the activity of preaching can be considered as an alternative way.

Omar, et al. [7] conducted a study about the role of social media in da'wah activity in Malaysia. In this study, it was concluded that the use of social media is necessary in preaching activity. The popularity of social media could be an alternative way to spread islamic messages to the people effectively. In that way, the scholars need to gain an ability to use social media as a channel of preaching.

All of the literature above have stated that information technology could be an alternative way to spread islamic knowledge. However, based on our best knowledge, the study that explore the relationship between religion preaching and Twitter is rare. As a matter of fact, Twitter becomes new phenomena in preaching community especially in Indonesia. Moreover, there are various interesting topics that need to be revealed. That is what this study aiming.

3. Methods

3.1. Questionnaire Analysis

This research is positivist in nature. It focuses on the influence of religion-related tweets posted by islamic preacher on Twitter activity of their follower. It used qualitative and quantitative analysis method. Research data are retrieved from questionnaire. The survey was conducted in July 2016. The participants were an active user ranging from 18 to 30 years old in age. We employ random sampling and also snowball sampling technique to choose participants.

After that, the data we retrieved from survey was analyzed. We use correlation bivariate analysis to inquire the influence of social media on follower's activity.

3.2. Tweet Extraction

We collected 11060 tweet data from five scholars who have the most followers from the respondents. Tweet dataset was grabbed by using the *TwitterSearch* library in Python. The tweet extraction task aims to obtain the topic of the tweets posted by the scholars. In particular, we conducted the clustering task using K-Means method.

Preceding the clustering task, we need to clean the Twitter messages by performing preprocessing tasks. The preprocessing tasks are removing URLs, removing tweet symbols, removing punctuations, case folding, tokenization, and removing stopword. The first task is removing URLs from the tweet. Secondly, tweet symbols such as @username, #hashtags, RT(retweet), etc. will be omitted in this step. The next step is removing punctuations from the tweet. Case folding is the process to convert words into same form, for instance lower case. After that, tokenization task will be conducted to divide the sentence into some parts called token. Finally, stopword removal is carried out by eliminating the common and frequent words which do not have the significant influence in the sentence. Figure 2 describes the detail steps in preprocessing task.

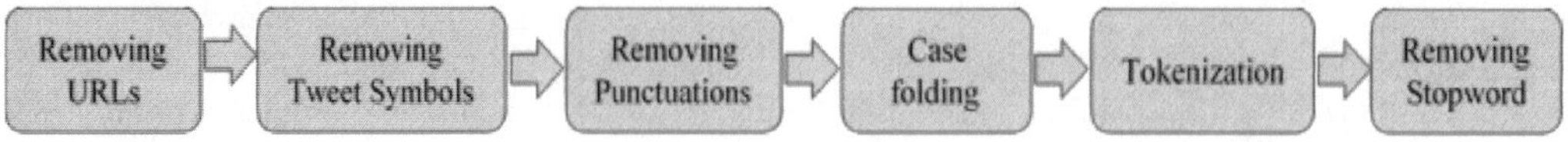

Figure 2. Preprocessing Task

4. Result and Discussion

In this study, the impact of social media is defined as the relationship between the frequency of kultwit given by scholars and the frequency of their follower read that kultwit. The number of tweet, to some extent, will attract other user to follow that tweet. Here, we defined two variable namely FD as independent variable and FM as dependent variable.

Variable 1 (FD): The frequency of tweet..............................(1)

Variable 2 (FM): The frequency of tweet follower..................(2)

The correlation analysis done with the two variable shows a strong relationship with coefficient 0.820 (r=0.820) in significant level of 0.01.

Table 1. Correlation Analysis

		frek_tweet	frek_follow
frek_tweet	Pearson Correlation	1	.820**
	Sig. (2-tailed)		.000
	N	102	102
frek_follow	Pearson Correlation	.820**	1
	Sig. (2-tailed)	.000	
	N	102	102

**. Correlation is significant at the 0.01 level (2-tailed).

Based on the data described in Table 3, it can be concluded that the number of tweets posted is strongly correlated with the number of follower that follow the tweets. It is somehow can be considered as the enthusiasm of the followers to know more about islamic messages contained in the tweets.

Furthermore, in order to dig more about the topic of the tweets, this study has carried out the clustering task using K-Means method. In this case, the *k* value for K-Means clustering is 5. Additionally, we also assign the maximum iteration of K-Means with 1000 times iterations. To identify the topics of the tweets, we explored the top ten frequent terms for each cluster that can be seen in Figure 3.

```
Top terms per cluster:
Cluster 0:  hati  islam  baik  puasa  anak  hidup  shalat  baca  mudah  kebaikan
Cluster 1:  maksiat  taat  dosa  udahputusinaja  pacaran  hati  pasti  berbuat  allah  cinta
Cluster 2:  doa  mhn  semoga  shbtku  shalawat  allah  minta  alloh  ayoo  haji
Cluster 3:  alloh  kajian  silakan  maha  hati  hadir  ahad  tauhiid  istiqlal  agendaaa
Cluster 4:  allah  semoga  maha  cinta  baik  moga  islam  ridha  hati  kebaikan
```

Figure 3. Top Terms per Cluster

Based on the top terms per cluster above, it can be seen that each cluster has a specific subject. First, Cluster 0 discuss more about *manajemen hati* (heart management) and worship. The second cluster talks about sins, fidelity, and love among young people. The tweet contents in the third cluster are about supplication, prayer to Allah (God), and persuade the followers to do something. The fourth cluster contains more about the information to the followers about the lecture schedule, the invitation to people to come to the lectures, and the agenda of the scholars. The last cluster talks about purification of the soul *(tazkiyatun nafs)* and supplication.

5. Concluding Remarks and Future Works

Islamic preachers and scholars have been using Twitter as a mean to share about Islamic knowledge. Various topics were shared by means of special kind of tweet called *kultwit*. The way preachers share their information is broad in variety. There is a correlation between the number of tweets with the number of follower. The more preacher post a tweets the more follower they attained.

Moreover, it can be seen that there are some popular topics posted by the scholars. Based on the clustering result, several topics are identified as the most frequently occurred. The tweets which contain about worship, love, information, and purification of soul have become the most favourable subject among preachers and scholars.

Further study can be held on two topics. First, research can be done in analyzing the content of Twitter messages using natural language processing technique. Second, empirical study can be held to prove the validity of factors affecting user acceptance of religion-related tweets.

6. Acknowledgement
The authors would like to thank Department of Informatics at Universitas Islam Indonesia for technical and financial support.

References
[1] Hidayatullah A F 2016 *Twitter Sebagai Media Dakwah* Jurnal TEKNOIN **vol 22** issue 1
[2] Lovheim M 2008 Rethinking cyberreligion?: Teens, religion and the Internet in *Sweden Nordicom Review* **vol 29** issue 2 p 205-17.
[3] Hosseini S E and Ramchahi A A 2014 The Impact of Information Technology on Islamic Behaviour *Journal of Multidisciplinary Engineering Science and Technology (JMEST)* vol **1** issue 5 p 135-41.
[4] Saefulloh A 2014 Cyberdakwah sebagai Media Alternatif Dakwah *ISLAMICA: Jurnal Studi Keislaman* p 138-60.
[5] Adam F, Anuar M M and Ali A H 2014 The use of blog as a medium of Islamic da'wah in Malaysia *International Journal of Sustainable Human Development* p 74-80.
[6] Agboola A K 2014 Religious Communication and Internet Usage: How Do Muslims Fare on the Web? *Middle-East Journal of Scientific Research 20 (Language for Communication and Learning)* p 11-20.
[7] Omar F I, Hassan N A, and Sallehuddin I S 2015 Role of Social Media in Disseminating Dakwah *Islamic perspectives relating to business, arts, culture and communication* p 43-55.

An Expert System for Diagnosis of Sleep Disorder Using Fuzzy Rule-Based Classification Systems

Lala Septem Riza, Mila Pradini, Eka Fitrajaya Rahman, and Rasim
Department of Computer Science Education, Universitas Pendidikan Indonesia, Indonesia
E-mail: lala.s.riza@upi.edu, mokvitasari@gmail.com, efitrajaya14@yahoo.co.id, rasim@upi.edu

Abstract. Sleep disorder is an anomaly that could cause problems for someone' sleeping pattern. Nowadays, it becomes an issue since people are getting busy with their own business and have no time to visit the doctors. Therefore, this research aims to develop a system used for diagnosis of sleep disorder using Fuzzy Rule-Based Classification System (FRBCS). FRBCS is a method based on the fuzzy set concepts. It consists of two steps: (i) constructing a model/knowledge involving rulebase and database, and (ii) prediction over new data. In this case, the knowledge is obtained from experts whereas in the prediction stage, we perform fuzzification, inference, and classification. Then, a platform implementing the method is built with a combination between PHP and the R programming language using the "Shiny" package. To validate the system that has been made, some experiments have been done using data from a psychiatric hospital in West Java, Indonesia. Accuracy of the result and computation time are 84.85% and 0.0133 seconds, respectively.

1. Introduction

Sleep is human's basic need [1]. It is considered substantial because sleeping is a process when human's cellulars are generating new cells, regenerating damaged cells (i.e., natural healing mechanism), giving the organs time to rest, and also keeping body's metabolism and biochemistry process in balance [2]. Therefore, when someone does not have a good sleep, then he might has a sleep disorder. An example of this issue has been conducted by J. Kales and S. Healey in Los Angeles, and it results of the following data: 42,5% are anguishing insomnia, 11,2% are having nightmare, 7,1% are oversleep, 5,3% are delirious, and 2,5% are sleep walking [3].

Basically, sleep disorder happened in a long period could reduce cognitive capability, including less concentration, easy to forget, slow in thinking process, also could affect relationship between friends and families, change on biological sleep cycle, immune deficiencies, reduce work performance, easily offended, depression, exhaustion, and eventually could endanger self safety or any others [4]. Based on this explanation, it can be concluded that sleep disorder is a health problem that has to be solved.

In order to solve sleep disorder, mostly patients should consult in face to face with medical assistance [5]. Yet, it could inflict other problems because of time and distance. Therefore, alternative solutions should be proposed, which is by developing a consultation media that could be accessed by many people and not rely on time and distance. Expert systems are basically able to handle these issues. Moreover, the systems have been applied to many areas. For example, it was applied in the disaster management system, namely flood control and defenses [6]. We can also find other applications of the expert systems, such as in determining defence strategy in basketball [7], for estimating intrinsic susceptibility in fishing [8], and for detecting breast cancer [9], fever disease [10], hearth disease [11], and diabetes detection [12]. Regarding with solving for sleep disorder, we can find the following research: [13, 14]. Moreover, there are some studies related to solving sleep disorder by using machine-learning approaches, such as [15, 16].

In this research, we attempt to develop an expert system that capable to provide diagnostic possibility of someone suffering sleep disorder using FRBCS. It is a part of well known methods in soft computing based on fuzzy concepts [17] to solve classification problem [18]. It is capable to classify uncertainty value into classes of groups that had been determined before. Furthermore, FRBCS can generate a model with two ways: (i) expert systems where knowledge is determined by human experts, and (ii) machine-learning approach requesting data training for generating a model. In this case, we are using FRBCS as the expert-system model.

The remainder of this paper is structured as follows. Section 2 and 3 briefly introduces to sleep disorder and fuzzy sets along with FRBCS. In Section 4, we describe how to construct a model based on FRBCS and implement the model. Section 5 presents results and discussion of some experiments. Finally, Section 6 concludes the paper.

2. Sleep Disorder

Sleep disorder happens when someone is not experiencing NREM or REM cycle normally. Sleep disorder can be interpreted as a disorder that can cause problems in someone's sleep pattern, either because of wakeful, frequent waking at night, or incapability to back to sleep after awakens [19].

There are two types of sleep disorder disease, which are organic type and non-organic type (functional). Organic sleep disorder type can be observed by changes in someone's nerve. Whilst, functional sleep disorder can be observed by someone's psychiatric and behavior. In this research, we are focusing on functional sleep disorder, which is divided into 4 types of sleep disorder: insomnia, depression, anxious, and psychotic [19]. The following is an explanation about these type of sleep disorder:

(i) Insomnia: Its common symptom is inability to sleep or unable to sleep well. Including getting up early in the morning. Basically, insomnia embraces so many this, such as hard to sleep well and even someone who awake from sleep but still have not had enough sleep can also be included as insomnia [20].

(ii) Depression: It is a bad mood which is occurring in a period of time. When someone suffering from depression, he will feel prolonged sad, desperate, doesn't have motivation to do any activity [21]. Basically, besides causing insomnia, depression also inflicts desire to sleep all the time because he want to run away from his problem. Depression is capable to cause insomnia and vice versa.

(iii) Anxious: Anxiety can be defined as an unpleasant emotional state, which is characterized by subjective feelings such as tension, anxiety, worry, and also characterized by active central nervous system [21]. Anxiety is also a bodily diseases and mental disorders, as well as highly interrelated. In psychological terms, this disorder is called Anxiety Disorders (AD). One possible cause is a chemical imbalance in the body, changes in brain structure, environmental, trauma, phobia, etc.

(iv) Psychotic: It is a mental disorder that characterized by the inability of individuals to assess the reality of the matter, so that people who have this disease have major symptoms of hallucinations, delusions and garbled speech [21].

Generally psychotic triggered by a person's inability to resolve the problem that is received by someone. How to control, treat, and respond to the problem is very different from one person to another, so that the level of severity of psychotic suffered very diverse although equally struck by the same problem.

3. Fuzzy Sets and Fuzzy Rule-Based Classification Systems

Fuzzy set theory proposed by Zadeh in 1965, as an extension of classical set theory to model sets whose elements have degrees of membership [17]. So, instead of having two values: a member and non-member, fuzzy set allows the degree of membership set, which is defined by a value between zero and one. The value of one means that an object is a member of the set, a value of zero means that it is not a member, and the value somewhere in-between shows the value of partial membership. This membership can be defined by membership functions.

Fuzzy set theory provides tools to effectively represent the concept of linguistic variables and rules, be natural to represent the model of a human expert knowledge. A key concept is that of the linguistic variables, defined as a variable whose value is a linguistic term, each with semantics described by fuzzy sets [22]. Value refers to the label to represent linguistic knowledge means determined by the degree of membership functions. For example, a_1 = "cold" with the level of $\mu = 0.8$ means that the variable a_1 has a linguistic value represented by the label "cold", whose meaning is determined by the degree of 0.8.

In this study, we use the trapezoidal membership function that can be expressed as follow:

$$\mu_{T(a,b,c,d)(X)} = \begin{cases} 0 & x \leq a \text{ or } x > c \\ \frac{x-a}{b-a} & 0 < x \leq b \\ 1 & b < x \leq c \\ \frac{x-d}{c-d} & c < x \leq d \end{cases}$$

And for the operation T-Norm, we use the following expression:

$$\mu_{A \cap B}(X) = min(\mu_A(X), \mu_B(X)),$$

while S-Norm, the following equation is used:

$$\mu_{A \cup B}(X) = max(\mu_A(X), \mu_B(X)).$$

FRBCS is a variant of fuzzy rule-based systems, which is used for handling classification tasks. The main characteristic of classification is that the consequents are in the form of a class or category. Therefore, in this type of model we can divide the symptoms section in the form of linguistic variables, and the consequent change to become part of the class C_j of a specified class $C = C_1, \ldots, C_M$. The simplest form of fuzzy rules introduced by Chi et. al. in 1996 was built with the class at the consequent [23]:

$$\textbf{IF } X_1 \text{ is } A_1 \text{ and } \ldots \text{ and } X_n \text{ is } A_n \textbf{ THEN } Y \text{ is } C_1,$$

4. Model Construction and Software Implementation for Sleep Disorder

In this section, a model is constructed by using FRBCS as illustrated in Figure 1. It can be seen that basically the processes can be divided into two parts: model construction and prediction over new data. First, an FRBCS model, known as knowledge, contains rulebase and database. Rulebase is a set of fuzzy rules while database contains label definitions and membership functions. Then, we predict new data by considering the obtained model and processes on FRBCS.

In fact, there are two strategies to build a model, which are by learning from data and by consulting with experts. In this case, we focus on developing a model by consulting with human experts. Therefore, the following are processes to obtain a model by consulting experts:

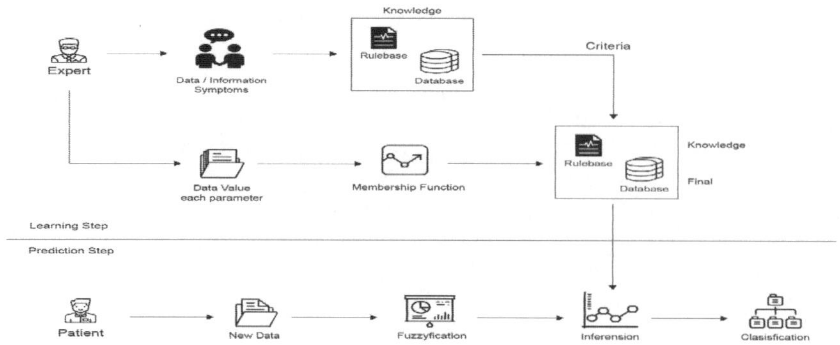

Figure 1. Model development for sleep disorder.

(i) To determine influencing symptoms: After studying literatures and consulting experts, we define 12 symptoms, as follows: physical disorder (G01), less concentration (G02), mood changing (G03), awakened frequency (G04), nightmare frequency (G05), hallucinations (G06), supposition (G07), talk chaotic (G08), diet changing (G09), moody (G10), hopeless (G11), and agitated (G12).

(ii) To determine database: In this step experts firstly need to define fuzzy labels on each symptom. In this case, we only consider two labels: "low" and "high", where the observation takes between 0 and 7 days. Therefore, we can draw the following membership functions for all symptoms:

$$\mu_{(low)}(X) = \begin{cases} 1 & x \leq 2 \\ \frac{5-x}{4} & 2 < x < 5 \\ 0 & x \geq 5 \end{cases} \qquad mu_{(high)}(X) = \begin{cases} 0 & x \leq 2 \\ \frac{x-2}{3} & 2 < x < 5 \\ 1 & x \geq 5 \end{cases}$$

(iii) To determine rulebase: A set of rules is obtained by consulting with experts. In this case, there are 392 rules collected in this step. These rules have represented four types of sleep disorders: insomnia (P01), depression (P02), anxious (P03), and psychotic (P04). For example, we have the following rules:

$R1$: **IF** $G01$ is *high* and $G02$ is *high* and $G03$ is *high* and $G04$ is *high* **THEN** $P01$

$R2$: **IF** $G01$ is *high* and $G02$ is *high* and $G03$ is *low* and $G04$ is *low* and $G09$ is *low* and $G10$ is *high* and $G11$ is *high* **THEN** $P02$

So, it can be seen that the main advantage by using FRBCS is that rules representing knowledge are in fuzzy terms so that naturally we can understand easily.

(iv) To reduce and optimize the rulebase: Since there are more than 100 fuzzy rules generated by experts, we have to reduce them by considering the following criteria: deleting redundant rules, deleting minor rules conflicted with the major ones, and choosing the simpler rules. From this step, we finally obtain 42 fuzzy rules.

So, by following four steps above, we obtain knowledge representing the FRBCS model containing rulebase and database. Then, this model will be used to predict new data.

For prediction, actually we just follow common procedures in FRBCS as explained in [18, 23, 24]. In this stage, we need to perform fuzzification, inference, and classification. In fuzzification, converting real values into fuzzy labels, and then their degrees of membership

Table 1. Results: 33 samples of patients

No	G01	G02	G03	G04	G05	G06	G07	G08	G09	G10	G11	G12	Pred	Actual
1	6	7	7	7	5	5	6	5	7	7	7	0	P01	P04
2	5	6	5	4	0	1	2	0	5	7	5	0	P02	P02
3	7	4	5	6	0	2	1	0	4	1	2	7	P03	P03
...	...	...	...	...	...	...	...	...	...	...	...	...	...	...
31	7	7	1	1	1	1	1	1	1	1	1	1	Neg	Neg
32	4	0	0	0	5	0	0	0	0	0	0	2	Neg	Neg
33	5	6	4	0	0	0	0	0	4	6	0	0	P02	P02

function is calculated. Inference refers to determine weight according to the operator T-norm and S-norm, so that we know which rules will be fired. Based on inference, then we decide which decision should be taken.

After designing the methodology for solving the problems, we implement it into system by using the "shiny" package [25] that connect two languages: PHP and R. It means we can basically access R through the web GUI as illustrated in Figure 2.

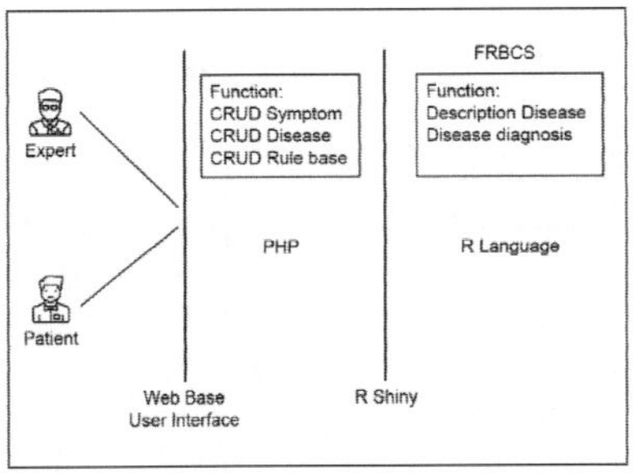

Figure 2. Architecture of the system.

5. Results and Discussion

The data used for testing are 33 samples obtained from one of the Psychiatric Hospital in West Java. The data consist of experienced symptoms that may be diagnosed as sleep disorder.

Table 1 shows some patients who have some experiences on related symptoms and their predicted and actual diagnosis. The symptoms, which are $G01$, $G02$, ..., $G012$, have values between 0 to 7 days. Diagnosis of the samples are $P01$, $P02$, $P03$, $P04$, and Neg representing insomnia, depression, anxious, psychotic, and negative for all diseases.

From these results, we obtain the accuracy of the experiments, which is 84.85% with the average of computation cost is 0.0133 seconds. In more detail, we have calculated confusion matrix on predicted and actual values. From this analysis, we obtained several measurements, as follows: $1.277e-10$ for P-value, 0.8099 for Kappa, Sensitivity of each class (i.e., 0.8333, 0.7, 1, 0.8571, and 1 for Neg, $P01$, $P02$, $P03$, and $P04$, respectively), and Specificity of each class (i.e., 1, 1, 0.9259, 0.9615, and 0.9310 for Neg, $P01$, $P02$, $P03$, and $P04$, respectively). Basically, it means that the proposed method can be used to predict sleep disorder effectively.

6. Conclusion and Future Work

The research can be summarized as follows:

(i) We have provided a model design to predict diagnosis of sleep disorder by using fuzzy rule-based classification systems.

(ii) According to the experiments, we obtained a reasonable result, with the accuracy of 84.85%. Moreover, the average of the computation time is 0.0133 seconds.

As future work, we plan to extend the diagnosis of sleep order by using Rough Set Theory [26].

References

[1] Honkus V L 2003 *Critical care nursing quarterly* **26** 179-91
[2] Braverman E R, Pfeifer C C, Blum K and Smayda R 2003 *The healing nutrients within: facts, findings, and new research on amino acids* (Basic Health Publications, Inc.)
[3] Kales J D and Healey S 1979 Am J *Psychiatry* **36** 1257-62
[4] Ancoli-Israel S and Cooke J R 2005 *Journal of the American Geriatrics Society* 53 S264-S271
[5] Ford D E and Kamerow D B 1989 *Jama* **262** 1479-84
[6] Cuena J 1983 *IJCAI* (Citeseer) pp 246-49
[7] Lin C C, Chen V, Yu C C, Lin Y C et al. 2006 *Proceedings of the 7th WSEAS International Conference on Fuzzy Systems* (World Scientific and Engineering Academy and Society (WSEAS)) pp 49-54
[8] Cheung W W, Pitcher T J and Pauly D 2005 *Biological conservation* **124** 97-111
[9] Karabatak M and Ince M C 2009 *Expert systems with Applications* **36** 3465-69
[10] Mulyani Y, Rahman E F, Herbert and Riza L S 2016 *International Conference on Science in Information Technology* (ICSITech) pp 367-71
[11] Turkoglu I, Arslan A and Ilkay E 2002 *Expert Systems with Applications* **23** 229-36
[12] Polat K and Güneş S 2007 *Digital Signal Processing* **17** 702-10
[13] Yildiz A, Akın M and Poyraz M 2011 *Expert Systems with Applications* **38** 12880-90
[14] Ălvarez-Estévez D and Moret-Bonillo V 2009 4th *European Conference of the International Federation for Medical and Biological Engineering* (Springer) pp 1121-24
[15] Wongsirichot T and Hanskunatai A 2015 *International Conference on Intelligent Computing* (Springer) pp 510-21
[16] Behar J, Roebuck A, Shahid M, Daly J, Hallack A, Palmius N, Stradling J and Cliford G D 2015 *IEEE journal of biomedical and health informatics* **19** 325-31
[17] Zadeh L A 1965 *Information and control* **8** 338-53
[18] Riza L, Bergmeir C, Herrera F and Benítez J 2015 *Journal of Statistical Software* 65 1-30
[19] American Academy of Sleep Medicine and others 2005 *The international classification of sleep disorders: diagnostic and coding manual* (American Acad. of Sleep Medicine)
[20] Roth T 2007 *Journal of clinical sleep medicine: JCSM: oficial publication of the American Academy of Sleep Medicine* **3** S7
[21] Goodwin F K and Jamison K R 2007 *Manic-depressive illness: bipolar disorders and recurrent depression* vol 1 (Oxford University Press)
[22] Zadeh L A 1975 *Information sciences* **8** 199-249
[23] Chi Z, Yan H and Pham T 1996 *Fuzzy algorithms: with applications to image processing and pattern recognition* vol 10 (World Scientific)
[24] Riza L S, Bergmeir C, Herrera F and Benítez J M 2014 2014 *IEEE International Conference on Fuzzy Systems* (FUZZ-IEEE) pp 2149-55
[25] Chang W, Cheng J, Allaire J, Xie Y and McPherson J URL http://CRAN. R-project. org/package= shiny. R package version 0.11
[26] Riza L S, Janusz A, Bergmeir C, Cornelis C, Herrera F, Ślezak D and Benítez J M 2014 Information Sciences **287** 68-89

Prediction of Disease Case Severity Level To Determine INA CBGs Rate

Sukma Puspitorini[1], Sri Kusumadewi[2], Linda Rosita[3]

[1]Informatics Engineering, STMIK Nurdin Hamzah, Kota Jambi, Indonesia
[2]Faculty of Industrial Technology, Universitas Islam Indonesia, Yogyakarta, Indonesia
[3]Faculty of Medicine, Universitas Islam Indonesia, Yogyakarta, Indonesia

sukm4pit@gmail.com, cicie@uii.ac.id

Abstract. Indonesian Case-Based Groups (INA CBGs) is case-mix payment system using software grouper application. INA CBGs consisting of four digits code where the last digits indicating the severity level of disease cases. Severity level influence by secondary diagnosis (complications and co-morbidity) related to resource intensity level. It is medical resources used to treat a hospitalized patient. Objectives of this research is developing decision support system to predict severity level of disease cases and illustrate INA CBGs rate by using data mining decision tree classification model. Primary diagnosis (DU), first secondary diagnosis (DS 1), and second secondary diagnosis (DS 2) are attributes that used as input of severity level. The training process using C4.5 algorithm and the rules will represent in the IF-THEN form. Credibility of the system analyzed through testing process and confusion matrix present the results. Outcome of this research shows that first secondary diagnosis influence significant to form severity level predicting rules from new disease cases and INA CBGs rate illustration.

1. Background and Objective

Case - mix is a disease classification system that combines similar or same clinical characteristic disease, with the cost of medical treatment in a hospital that generally required same resources and treatment costs. Diagnosis grouping determined based on two principles, clinical homogeneity (patients who have clinical similarity) and resource homogeneity (patients using same sources intensity for similar therapy/resource consumption). INA CBGs (Indonesian Case-Based Groups) is case-mix payment system using software grouper application to determine the rates of health care facilities in accordance PMK No.69 Year 2013 about the Standard Rates of Health Services [1]. INA CBGs rates determined by an INA CBGs code consisting of four digits financing components write in alphanumeric code. Each digit explain case-mix main group (CMG), case-type group, case-based group (CBG), and severity level. CMG related with organ system that classify into 31 group, case-type group indicate 9 type of treatment and procedure type, and CBG refer to specific case-based group denoted with numeric from 01 to 99 [2]. Severity level (SL) as the last code that indicates disease cases severity associated with a secondary diagnosis that can prolong hospitalized patient's length of stay (LOS) and caused rates varies on each case-mix main group (CMG).

The level of severity affect by complications and co-morbidity. Hospitalize patient grouped into 3 categories level of severity. Level I (mild) is diagnosis without complications and co-morbidity. Level II (medium) is diagnosis with mild complications and co-morbidity, and level III (severe) is diagnosis with major complications and co-morbidity [3]. If severity level could predict first during the treatment process, INA CBGs operators could anticipate increasing severity level during the treatment period and give illustrasion of INA CBGs rates before grouping process using INA CBGs software grouper.

In this study, severity levels determination relied on data grouping report by INA CBGs application. Result of the report is a large amount of data bases containing information and medical data or non-medical patients which analyzed to determine the pattern of prediction rules determining the severity level. Based on these, building decision support system (DSS) may help INA CBGs operators predicting severity level from disease diagnosis by knowing the patterns of patient data and classified using data mining techniques. Data mining is one of captivating soft computing techniques that widely use to analyze the hidden patterns in large amounts of information [3]. The rules of severity level determine by the classification model using C4.5 decision tree algorithm. Purpose of this research is implementing data mining decision tree classification model to develop severity level of disease case prediction and overview the rates. The C4.5 algorithm uses as decision tree classification. Attributes processed are age, primary diagnosis, and secondary diagnoses in hospitalized patients with hypertension, stroke, and diabetes mellitus type 2.

2. Methods

The following research analyzing 1000 patients from two government hospitals (RSUD Raden Mataher in Jambi and RSUD Dr. R. Soeprapto in Cepu) who had primary diagnosis of hypertension, stroke, and diabetes mellitus type 2 (non-insulin dependent diabetes mellitus) as seen in Table 1. This study focused on patient that had primary diagnosis with two secondary diagnosis to reduce the complexity in decision tree forming. First secondary diagnosis consist of 160 disease code while second secondary diagnosis consist of 109 disease code. Row data cannot directly use and needs transformation to categorize attribute. As the result, DS 1 categorized into 77 blocks of diseases and DS 2 categorized into 58 blocks of diseases. Secondary diagnosis categorized based on diseases block refer to ICD-10 chapter.

Tabel 1. Primary Diagnosis

Block	Code	Disease Name
Hypertension	I10	Essential (Primary) Hypertension
	I119	Hypertensive Heart Disease Without (Congestive) Heart Failure
Diabetes Militus Type 2	E110	Non-Insulin-Dependent Diabetes Mellitus With Coma
	E111	Non-Insulin-Dependent Diabetes Mellitus With Ketoacidosis
	E117	Non-Insulin-Dependent Diabetes Mellitus With Multiple Complications
	E118	Non-Insulin-Dependent Diabetes Mellitus With Unspecified Complications
	E119	Non-Insulin-Dependent Diabetes Mellitus Without Complications
Stroke	I64	Stroke, Not Specified As Haemorrage Or Infarction

The severity level prediction rules represented in the form of "IF-THEN" rules. This method does not require prior knowledge management and be able to solve the simple case which has dimensions of big data [4]. However, the constraint in building a decision tree is an attribute or a variable which will be at the root of the tree. One of the methods to determining the root of the decision tree is C4.5 algorithm. The roots of trees calculated using the concept of entropy-gain [5]. Attribute with the highest gain value will be elected as the root of the tree. Entropy compute using equation (1) below.

$$Entropy(S) = \sum\nolimits_{t=1}^{n} -pi * log_2(pi) \tag{1}$$

While the gain is calculated by equation (2) below

$$Gain(S,A) = Entropy(S) - \sum\nolimits_{t=1}^{n} \frac{|S_i|}{|S|} * Entropy(S_i) \tag{2}$$

Where

S = the reference node
A = attribute
n = number of attributes partitions A
| Si | = Number of cases in the partition i and | S | = Number of cases in S
pi = proportion of Si to S and log2 pi = log/log2 .

In addition, to find out INA CBGs rates, patient historical data use to determine CMG and CBG code. Data analyze using data mining tools WEKA 3.7.4 with J48 decision tree algorithm classification model. Testing process explain that decision tree generate tree rules of CMG and CBG code. Case-type in INA CBGs code describe treatment in hospitalize or outpatients, with or without medical procedure. There are 9 case types of cases and this study analyzing hospitalize patients without procedure with default code '4'.

3. Study Design

Classification worked to assess specific object from a number of available classes [6]. In data mining classification models, historical data could used as training data, to gain knowledge (such as a description or prediction). The data set divided into training set and testing set. Training set uses to build classification model while testing set uses to figure model accuracy. Decision tree classification model is supervises learning that proper to make predictions on this research [4]. Stages of this study are : 1) Data collection. This stage done through field studies from hospital and interviewing medical records staff to get information related INA CBGs code and severity level. 2) Data selection. Research focused on patient that had hypertension, diabetes mellitus type 2, and stroke primary diagnosis with two second diagnosis to reduce the complexity in decision tree forming. 3) Data analyzing. Result of this process setting primary diagnosis (DU), age, first secondary diagnosis (DS 1), and second secondary diagnosis (DS 2) as an input attribute. 4) Analysis data mining models using C4.5 decision tree classification algorithm to built prototype of decision making model. It will classify training set and predicted severity level from new cases. 5) Model implementation by building web-based decision support system applications to train data and performed testing. 6) Testing process to check credibility and validity of the system.

4. Decision Modelling

Decision modelling necessary to ease severity level determination. Model is predictive using decision tree classification model. There are several process to model the decision that is 1) Data input process. Data input is in csv format and user could export it from excel file or manually input it from application. 2) Set training data. in this study, all input data will set as training data. 3) Building decision tree using C4.5 algorithm. Decision tree will generate rule to classify severity level based on primary diagnosis, age, and disease block of secondary diagnosis. 4) Data testing. This process conducted to determine validity from generated rule comparing with severity level result from INA CBGs grouping. 5) Credibility testing. System performance assess using confusion matrix.

5. Result

This application only serves as decision support software. Patient data that will predict has through the process of disease coding by medical record staff. User need entering medical record data like age, disease code of primary diagnosis (DU) and secondary diagnosis (DS 1 and DS 2) to determine severity level and rate prediction. Suppose user entering new input case. Patient age is 61 years with primary diagnosis code is I64, secondary diagnosis code are G21 and N309. Figure 1 shows the prediction result and information.

Kode Pasien	: 3
Umur	: 61
Kategori	: Lansia
Diagnosa Utama	: I64
Nama Penyakit	: Stroke, not specified as haemorrhage or infarction
Diagnosa Sekunder 1	
Kode ICD 10	: G21
Nama Penyakit	: Secondary parkinsonism
Blok Penyakit	: Kelainan extrapyramid dan gerakan
Diagnosa Sekunder 2	
Kode ICD 10	: N309
Nama Penyakit	: Cystitis, unspecified
Blok Penyakit	: Kelainan-kelainan lain sistem perkemihan
Severity Level	: SL 2

Figure 1. Prediction Information Result

Display information explain that patient with primary diagnosis *stroke that not specified as hemorrhage or infarction* with complication/co-morbidity *secondary parkinsonism* and *cystitis unspecified* has severity level II (SL 2) or medium severity level of disease case. Prediction of rate determine by clicking KLAIM button at the bottom and user will be redirected to page. Pres enter to assign CMG-CBG code. System classify new case into CMG-E and CBG-10. Then, select hospital type to determine regional and the type of the hospital. Suppose, user choose treatment in class 2. The result could see in Figure 2 below.

Kode INA CBGs	: E-4-10-I
Deskripsi	: PENYAKIT KENCING MANIS & GANGGUAN NUTRISI/ METABOLIK RINGAN
Kode CMG	: E
Tipe RS	: Regional 1 RS Kelas B
Kelas Rawat	: 2
Tarif	: Rp 3671353

Figure 2. INA CBGs Rates Prediction Result

System generate INA CBGs code E-4-10-I which mean that patient disease case classify as *mild diabetes and metabolic nutritional disorder* case-mix main group with specific CBG number 10. Patient hospitalized with no medical procedure treatment. Hospital region and type also influence INA CBGs rate.

Application testing performed to evaluate whether the severity level generated by the application fit with INA CBGs grouping result. The testing results from new cases in Table 2 below shows that from 15 cases, 13 cases predicted correctly while case 1, case 4, and case 5 has difference severity level result. This deviation is not surprising because there are partitions on the attributes that cannot determine the severity level class due to lack of base case. As the result, system classified severity level based on modus.

Tabel 2. Severity level prediction result using new data

Case	Age	DU	DS 1	DS 2	DSS SL	INA CBGs SL	INA CBG CODE	RATE REG 1 TYPE B (IDR)	RATE REG 3 TYPE C (IDR)
1	40	E110	A419		2	3	E-4-10-II	4.235.205	2.798.255
2	47	E117	N189	C539	1	1	E-4-10-I	3.059.460	1.820.720
3	50	E118	I500		2	2	E-4-10-II	4.235.205	2.798.255
4	52	E119	A162		1	2	E-4-10-I	3.059.460	1.820.720
5	67	E119	L309		2	1	E-4-10-II	4.235.205	2.798.255
6	38	I10	Z038		1	1	I-4-17-I	3.502.018	2.928.961
7	78	I10	R51		1	I	I-4-17-I	3.502.018	2.928.961
8	46	I10	G819		2	2	I-4-17-II	4.747.470	3.970.611
9	57	I10	R42		1	1	I-4-17-I	3.502.018	2.928.961
10	52	I10	J459	N309	2	2	I-4-17-II	4.747.470	3.970.611
11	79	I10	I500	E119	2	2	I-4-17-II	4.747.470	3.970.611
12	50	I10	G459		2	2	I-4-17-II	4.747.470	3.970.611
13	44	I119	K297		1	1	I-4-17-I	3.502.018	2.928.961
14	71	I64	I10		1	1	G-4-15-I	2.872.210	1.967.992
15	57	I64	E148		1	I	G-4-15-I	2.872.210	1.967.992

INA CBGs rate is different in every regional. It depend on hospital type and class of tratment. Table 2 illustrate prediction if disease case occure on heathcare facilities type B at regional 1 and type C at regional 3 with the same class treatment. Although the INA CBGs code is same but the rate has differences.

6. Conclusion

Conclusion from this study is that data mining models classification with decision tree could use to compose severity level of disease cases rule. The first secondary diagnosis is attributes that significant determining severity level of disease cases. It is precondition that checked first before another attributes. Differences in severity level at any cases of disease caused differences in INA CBGs rates. Several disease case could have same INA CBGs code but the rate is different. It happen because hospitals regional differences and different class of treatment.

7. Refference

[1] Departemen Kesehatan Republik Indonesia 2013 *Peraturan Menteri Kesehatan Republik Indonesia Nomor 69 Tahun 2013 Tentang Standar Tarif Pelayanan Kesehatan Pada Fasilitas Kesehatan Tingkat Pertama Dan Fasilitas Kesehatan Tingkat Lanjutan Dalam Penyelenggaraan Program Jaminan Kesehatan* http://www.depkes.go.id/resources/download/general/PMK No. 69 Th 2013 ttg Tarif Pelayanan Kesehatan Program JKN.pdf

[2] Kementrian Kesehatan Republik Indonesia 2014 *Peraturan Menteri Kesehatan Republik Indonesia No.27 Tahun 2014 Tentang Petunjuk Teknis Sistem Indonesian Case Base Groups*

(INA-CBGs) http://www.jkn.kemkes.go.id/attachment/unduhan/PMK No. 20 ttg Juknis Sistem INA CBGs.pdf

[3] Tim Coding National Casemix Center (NCC) Departemen Kesehatan RI 2012 *Koding INA CBGs* http://bppsdmk.depkes.go.id/ckfinder/userfiles/files/KODING INA CBG.pdf

[4] Widodo P P, *et al* 2013 *Penerapan Data Mining Dengan Matlab* (Bandung: Penerbit Rekayasa Sains).

[5] Kusumadewi S, *et al* 2009 *Informatika Kesehatan* (Yogyakarta: Graha Ilmu).

[6] Prasetyo E 2012 *Data Mining : Konsep Dan Aplikasi Menggunakan Matlab* (Yogyakarta: Andi Offset).

Software Size Estimation Using Activity Point

S Densumite[1] and P Muenchaisri[2]

[1, 2] Chulalongkorn University, Bangkok 10330, Thailand

Email: [1] sharif.d@student.chula.ac.th, [2] pornsiri.mu@chula.ac.th

Abstract. Software size is widely recognized as an important parameter for effort and cost estimation. Currently there are many methods for measuring software size including Source Line of Code (SLOC), Function Points (FP), Netherlands Software Metrics Users Association (NESMA), Common Software Measurement International Consortium (COSMIC), and Use Case Points (UCP). SLOC is physically counted after the software is developed. Other methods compute size from functional, technical, and/or environment aspects at early phase of software development. In this research, activity point approach is proposed to be another software size estimation method. Activity point is computed using activity diagram and adjusted with technical complexity factors (TCF), environment complexity factors (ECF), and people risk factors (PRF). An evaluation of the approach is present.

1. Introduction

Software size has been a challenge in software project management as an important parameter for effort and cost estimation. Software Line of Code (SLOC) is a software sizing physically counted after the software is developed. Many methods computed at early software development activities.

Allan Albrecht [1, 2] introduces Function Point Analysis (FPA) method to estimate software size in early phase which decomposes the complex system into simple systems. Unadjusted function point is adjusted with value adjustment factor (VAF) to represent software size as function points unit. Thirty years later, FPA has evolved into many methods such as IFPUG [3], NESMA [4] and COSMIC-FFP [5, 6]. Unified Modeling Language (UML) based object oriented metrics are widely used to create sizing model that inspired from function point including use case point (UCP) [7], sequence point (SP) [8], and class point (CP) [9].

UCP is a software size estimation approach in the early phase which uses transactions, scenarios, and actors to be main factors computed to software size, but it is not considered in use case relations that are directly affected to software size. This paper presents an approach to utilize activity diagrams named Activity Point (AP) which apply UCP in transactions, scenarios and use case relation to compute software size.

This paper is divided into 6 sections. Section 2 introduces related work. Section 3 provides an approach to calculate software size with AP approach. Section 4 describes a method to define weight factors. Section 5 presents an evaluation of the model. Finally, section 6 shows conclusions and future works.

2. Related work

A. Albrecht [1, 2] proposes Function Point Analysis (FPA) method to predict software size from functionalities which are composed of five components, including Internal Logical Files (ILF),

External Interface Files (EIF), External Inputs (EI), External Outputs (EO), and External Inquiry (EQ). Counting elements are multiplied with numerical rating and summed to give Unadjusted Function Point (UFP). UFP is adjusted by Technical Complexity Factor (TCF) to collect the result in function point unit size.

CFP approach presented in [5, 6] is applied to business application, real-time, and industrial software. The CFP approach is inspired from FPA and is mainly focused on data movement, which is classified into four types: entry, exit, read, and write. Software size will be replaced with the total number of data movements.

G Karner [7] proposes a model to measure software size using use case diagram named Use Case Point (UCP). UCP approach emphasizes in user interaction and use case transactions of each use case. UCP consists of two components: unadjusted actor point (UAP), and unadjusted use case weight (UUCW). The components are summed and adjusted with Technical Complexity Factor (TCF) and Environment Complexity Factor (ECF).

A. Sellami et al [8] use a set of procedures for sizing sequence diagrams at a high level of granularity using the COSMIC rules, and for the measurement of the structural size of sequence diagrams at a finer level of granularity. The approach uses three interactions in sequence diagram including Alternative (alt), which represents optional behavior that will be compiled in the system, Optional (op) operator that represents a behavioral choice that will be chosen follows the condition.

G. Costagliola [9] et al present sizing approach for object oriented products using class diagram called Class Point (CP). The approach has three steps. First, identifying and classifying the classes into four types namely problem domain, human interaction, data management, and task management type. Second, assigning a complexity level and weighting to each class. Finally, determining class point by adjusting the weighted sum from the previous step.

L Lavazza et al [10] improve UCP method using Path (P) and Transaction (T) which is described in the use case description to create a measurement model established by linear regression. G Robiolo et al [11] offer an alternative way by creating an effort model with linear regression using use case transactions and scenarios as main factors.

H. Madhavji et al [12] present quality factors which are impacted to software architecture in people issues, including procrastination, missing/late meetings, team members(s) not delivering sufficient work, poor planning, and group strategy.

3. Activity and use case diagram in two perspectives

The approach divides activities into two perspectives: high level and low level. High level of activity diagram presents use case's relationships that are directly affected to software size such as "glue code" which includes activities or incidents. On the other hand, low level of activity diagram describes transactions and scenarios that are separately computed and weight with ideal number which includes activities or incident.

Fig. 1 describes activity perspectives that are designed from use case diagram. There are transaction perspective and use case relation perspective presented in different level of abstraction as follows.

- Fig 1.a and 1.b show an activity diagram at low level that separately describes use case description of UC1 and UC2. Activity/Action represents transaction and Path represent scenario of use case diagram.
- Fig 1.c shows a high level activity diagram that describes use case's relation. Activity/Action represents each use case in use case diagram.

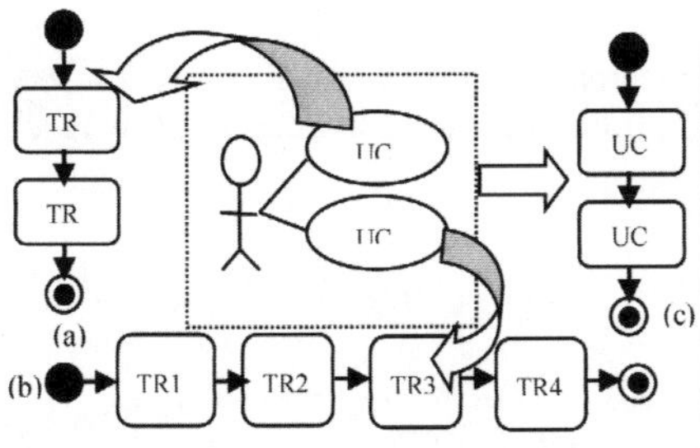

Figure 1. Activity perspectives.

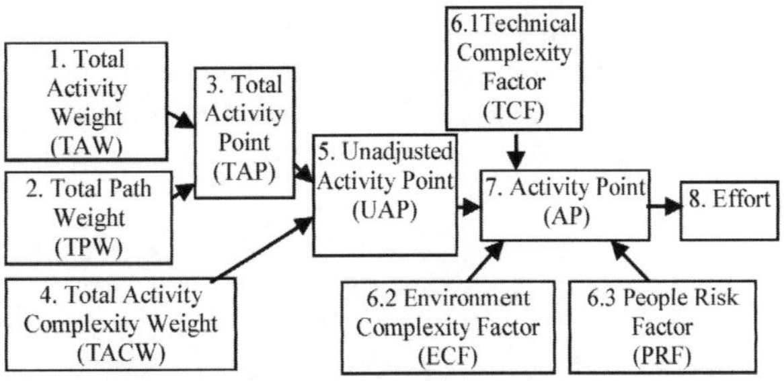

Figure 2. Overview of activity point approach.

4. Activity and use case diagram in two perspectives

In this section, Activity Point approach is proposed to estimate software size in early phase. Fig.2 describes Activity Point (AP) which is computed from Unadjusted Activity Point (UAP), Technical Complexity Factors (TCF), Environment Complexity Factors (ECF), and People Risk Factors (PRF). UAP is unadjusted size which is computed from Total Activity Point (TAP), and Total Activity Complexity Weight (TACW). Detail of each component is described in the following.

4.1. Total activity point

TAP is a factor, which presents size at low level, including TAW and TPW. The level of TAW and TPW is classified into 3 levels: simple, average, and complex. TAP is obtained by summation of TAW and TPW at row 1 in Table 1. A number of transactions in use case diagram is replaced by a number of activities or incidents. A number of scenarios is indicated by a number of paths in activity diagram. TAW is calculated from a Number of Activities or incidents (NA) level multiplied with weighting number at row 2 in Table 1. NA classification level and weight value are presented in Table 2. In the same way, TPW is computed from Number of Paths (NP) level multiplied with weighting number at row 3 Table 1. NP classification level and weight value are shown in Table 3.

4.2. Total activity complexity weight

TACW is a factor which represents a size at high level of activity diagram. The activity diagram will be transformed to Control Flow Graph (CFG) and specified complexity using McCabe Complexity approach. The total complexity represents size as Activity Complexity (AC). In addition, if the activity diagram consists of Folk and Join notation, there are possible cases as follows.

- Paths of fork/join are sequences. The paths will be combined to a single path as shown in Fig. 3 because it is equal to McCabe complexity result. The reason is parallel sequencing process does not affected to complexity increased.
- In contrast, Fig. 4 shows paths of fork/join including decision or loop in different path. The notations are connected to a single path that the notations sequence connected is insignificant meaning if the loop or decision comes first to calculating complexity.
- Synchronisation link between actions in different paths of fork/join is shown in Fig. 5 which the paths should be separated and linked together.

TACW is computed from AC and weighting number at row 4 in Table 1. AC is classified into three levels as simple, average, and complex which are shown in Table 4.

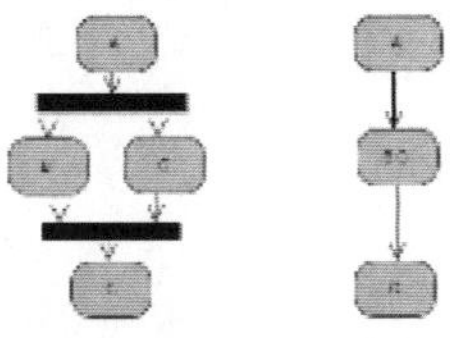

Figure 3. Sequencing fork/join transformation to control flow graph.

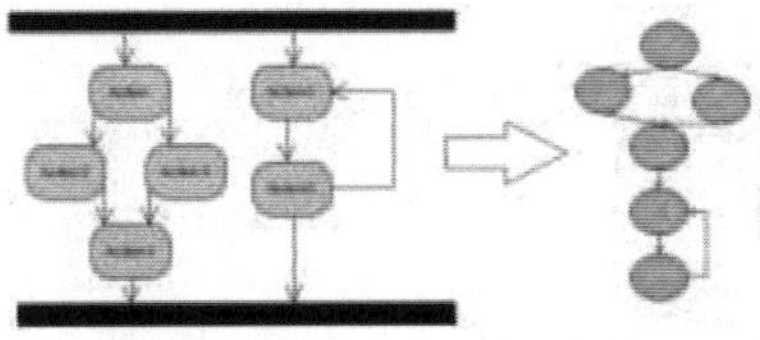

Figure 4. Complex fork/Join transformed to control flow graph.

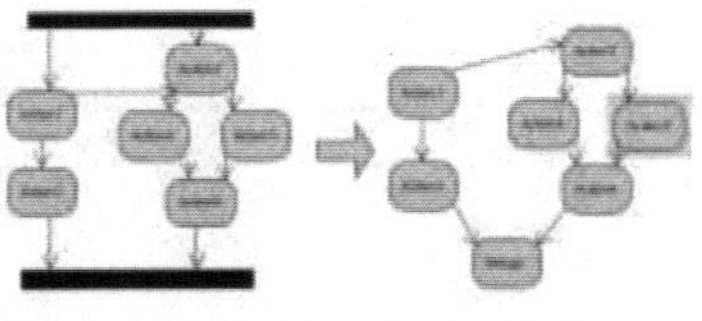

Figure 5. Synchronisation link perspectives.

4.3. *Unadjusted activity point*

UAP is an initial software size computed during software development from TAP and TACW at row 5 Table 1. The number of TAP (N) in the formula is equal to the total number of use cases at low level. TACW is described as size at high level.

4.4. *Activity point*

Activity point (AP) is a software development size, which is computed from UAP multiplied with adjusting factors at row 6 in Table 1.

The adjusting factors include TCF, ECF, and PRF which are subsequently shown in Table 5-7. TCF, ECF, and PRF are scored from 0 (not related) to 5 (highly related). TCF at row 7 in Table 1 is an adjusting factor in technical consideration. ECF at row 8 in Table 1 is an adjusting factor based on environment consideration. TCF and ECF calculation are the same as UCP approach in [7]. People risk factor (PRF) is an additional people factor proposed at row 9 in Table 1 which is used to find impact factors affected to AP. Formula at row 9 in Table 1 is created from estimated effort without PRF since PRF is assumed to be 1. The ideal number of PRF comes from the multiplier of the addition of personal factor (PF) and the ideal number that makes the lowest average MRE from 7 project training data sets which is presented at row 10 Table 1. PF in Table 7 is inspired from [12]. The highest risk is 5.2 when PF score is 35.

4.5. *Effort*

An estimation effort model is provided at row 11 in Table 1 which uses AP as size and is created by linear regression including predictive variable, and independent variable which subsequently represent actual effort and AP.

Table 1. Activity point equation.

No.	Equation
1	$TAP = TAW + TPW$
2	$TAW = \sum NA\ level * Weight$
3	$TPW = \sum NP\ level * Weight$
4	$TACW = \sum AC\ level * Weight$
5	$UAP = \sum_{i=1}^{N} TAP_i + TACW$
6	$AP = UAP * TCF * ECF * PRF$
7	$TCF = 0.6 + TF/100$
8	$ECF = 1.4 + (-0.03 * EF)$
9	$PRF = 1 + (0.12 * PF)$
10	$\text{Average MRE} = \frac{\sum_{i=1}^{N} (\frac{Actual\ Effort_i - Estimated\ Effort_i}{Estimated\ Effort_i})}{N}$
11	$Effort = a + b * AP$

Table 2. Activity/Action weight category.

NA Level	Description	Weight
Simple	Simple steps in activity or action between 1 and 3.	2
Average	More steps in activity or action between 4 and 7.	3
Complex	Much steps in activity or action that more than or equal 8	4

Table 3. Path weight category

NP Level	Description	Weight
Simple	Simple alternative path in activity between 1 and 3.	3
Average	More alternative path in activity between 4 and 7.	5
Complex	Much steps alternative path in activity between that more than or equal 8	9

Table 4. Activity complexity weight category.

AC Level	Description	Weight
Simple	Less complexity from McCabe Complexity is 1.	9
Average	Much Complexity from McCabe Complexity between 2 and 3.	14
High	High Complexity from McCabe Complexity more than or equal 4.	18

Table 5. Technical complexity factors.

Factor	Description	Weight
T1	Multi user experience and user interface design	2
T2	Response time/performance objectives	1
T3	Server configuration	1
T4	Requirement complexity	1
T5	Support multi devices	0.5
T6	Big data scaling support	0.5
T7	Multiple sites implement	1
T8	Portability to other platforms	2
T9	System maintenance	1
T10	Concurrent/parallel processing	1
T11	Security features	1
T12	Access for third parties	1
T13	End user training	1

Table 6. Environment complexity factors

Factor	Description	Weight
E1	Familiarity with development process used	1.5
E2	Application experience	0.5
E3	Object-oriented experience of team	1.0
E4	Lead analyst capability	0.5
E5	Motivation of the team	1.0
E6	Stability of requirements	2.0
E7	Part-time staff	-1.0
E8	Difficult programming language	-1.0

Table 7. People risk factors.

Factor	Description	Weight
R1	Team member(s) Procrastination.	2
R2	Missing/late meeting	0.5
R3	Team member(s) often not delivering adequate work.	2
R4	Poor planing and group strategy.	1
R5	Customer Orientation	1.5

5. Defining weight values

Weight values of factors in AP approach are calculated using the algorithm as shown in Fig. 6. The multiplier of NA level, NP level, and AC level in (2) - (4) are executed from training data sets of 7 projects including 4 enterprise software developed with JAVA, 2 mobile applications developed with JavaScript, and 1 e-commerce website developed with PHP. Number of activities/actions, paths, and complexity are calculated to find correlation with actual effort. The correlation indicates which factor is more affected to actual effort. The correlation of activities/actions, paths, and complexity with effort are subsequently resulted as 0.823, 0.844, and 0.665. It means NP is the most affected to actual effort comparing with others. Therefore, the proportionate percentage affecting factor with actual effort is 50 percent of TPW, 40 percent of TAW, and 10 percent of TACW. In Table 2 – 4, weight values are between 1 and 20. NA level, NP level, and AC level are weighted with the ideal number from the range which gives UAP the highest correlation with actual effort. UAP is a prior adjusted size that does not consider external adjusting factor. Therefore, adjusting factors should be considered as external factors to adjust UAP to be a final size as AP unit.

1 INPUT projects data include number of transactions, scenarios, activity complexity, and actual effort.
2 FIND correlation dependence with project's actual effort
3 SET percentage affected to effort of transactions, scenarios, and activity complexity from the correlation.
4 MULTIPLY actual effort with the percentage.
5 STORE representing transactions, scenarios, and activity complexity point of each project from previous step.
6 INITIAL preferred range of weight value.
7 FIND ideal weight value of each point that gives the closest to actual effort of each project with looping.
8 SET standard weight value by average ideal weight result

Figure 6. Algorithm for defining weight values

6. Evaluation

Figure 1 is a demonstrating diagram which shows activity diagrams at low level and high level. At low level, there are a simple activity level in (1.a) and an average level in (1.b). The path level of low level in (1.a) and (1.b) are simple. TAW of (1.a) is equal to 2 and (1.b) is equal to 3. TPW of (1.a) and (1.b) are equal to 3. Then, TAP of (1.a) and (1.b) are subsequently equal to 5 and 6. At high level, McCabe complexity of (1.c) is equal to 3 as average level. Thus, TACW is equal to 14. TAP of (1.a), (1.b), and TACW are summed to be UAP that is equal to 25. Assuming that adjusting factor score is equal to 1 that gives result of TCF, ECF, and PRF is 0.74, 1.265, and 1.7 consecutively. Therefore, AP is calculated from multiplier of UAP, and adjusting factor which gives result of 39.50 and uses as an input to estimate software development effort.

Three projects as test data sets including 2 enterprise software developed with JAVA, and 1 mobile application developed with JavaScript are used to evaluate the AP approach. The test data sets are used to evaluate AP model. Average Magnitude Relative error (MRE) and Prediction Quality (PRED) are presented as quality factors of predictive expected effort. Table 8 shows the number of paths of each project at low level. Table 9 presents the number of activity/action of each project at low level. Table 10 provides the number of activity complexity of each project at high level. Activity point, estimated effort, and actual effort are presented in Table 11 with the average MRE of 0.09. The average MRE is calculated from absolute RE of each project compared with actual effort and estimated effort. PRED(0.25) is resulted as 1 to present quality of AP which shows the percentage of project that MRE lower than 25% comes from 100% of total projects.

Table 8. Data of activity paths

Project	Quantity of Path Levels		
	simple	*average*	*complex*
1	6	6	3
2	4	3	2
3	16	5	0

Table 9. Data of activity / action.

Project	Quantity of activity/action levels		
	simple	*average*	*complex*
1	0	3	12
2	0	3	6
2	4	7	10

Table 10. Data of activity complexity.

Project	Quantity of Activity Complexity		
	simple	*average*	*complex*
1	0	0	1
2	1	0	0
3	0	0	1

Table 11. Effort evaluation.

Project	Activity Point	Estimated Effort	Actual Effort	MRE
1	341.65	286.19	301	0.05
2	77.47	84.09	102	0.21
3	89.65	93.41	92	0.02
Average MRE				0.09
PRED(0.25)				1

7. Conclusion

Software size estimation is a factor that many effort models use. This research considers in estimating effort using software size estimation based on activity diagram named Activity Point. The approach emphasizes metrics that are used in use case point approach and additional metrics proposed. The additional metrics focus on use case relation as activities or incidents, paths, and complexity in activity diagram. Unadjusted activity point is a prior adjusted size. A new adjusting factor named people risk factor is proposed. Technical Complexity Factor (TCF), and Environment Complexity Factors (ECF) are alike presented in use case point approach. Evaluation is presented with absolute mean relative error and PRED(25). The average MRE is calculated from absolute RE of three projects compare with actual effort and estimated effort which is resulted as 0.09. PRED(0.25) is resulted as 1.0 to present quality of AP which shows the percentage of project that MRE lower than 25% comes from 100% of total projects.

In the future works, the researchers will increase the efficiency of the model by working with more data sets, suitable weight value factors and adjusting factor scored criterion. A comparing with other size estimation approaches may be investigated. An automated tool supporting the proposed will be developed.

References

[1] Albrecht A J 1979 Measuring Application Development Productivity *Proceedings of the Joint SHARE, GUIDE, and IBM Application Development Symposium, Monterey, California, October 14–17* (IBM Corporation 1979) pp. 83–92.

[2] Kim SE,Lively W, and Simmons D 2006 An Effort Estimation by UML Points in Early Stage of Software Development *Software Engineering Research.*

[3] International Function Point Users Group (IFPUG) January 2010 *The Function Point Counting Practices Manual version 4.3.1* [EB/OL]. www.ifpug.org.

[4] ISO/IEC24750 2005 *Software engineering-NESMA functional size measurement method version 2.1-Definitions and counting guidelines for the application of Function Points Analysis[S]* (Geneva: International Organization for Standardization-ISO)

[5] ISO/IEC19761 2003 *Software Engineering-COSMIC-FFP A Functional Size Measurement Method[S]* (Geneva: International Organization for Standardization-ISO)

[6] Abdullah NAS, Rusli NIA, and Ibrahim MF 2013 A case study in COSMIC functional size measurement: Angry Bird Mobile Application in *Open Systems (ICOS) 2013 IEEE Conference* pp. 139-44.

[7] Karner G 1993 Resource Estimation for Objectory Projects *Objective Systems SF AB.*

[8] Sellami A, Hakim H, Abran A, and Ben-Abdallah H 2015 A measurement method for sizing the structure of UML sequence diagrams *Information and Software Technology* vol. **59** pp. 222-32.

[9] Costagliola G, Ferrucci F, Tortora G, and Vitiello G 2005 Class point: an approach for the size estimation of object-oriented systems *IEEE Transactions on Software Engineering* vol. **31** pp. 52-74.

[10] Lavazza L and Robiolo G 2010 Introducing the evaluation of complexity in functional size measurement: a UML-based approach *the Proceedings of the 2010 ACMIEEE International Symposium on Empirical Software Engineering and Measurement* (Bolzano-Bozen, Italy).

[11] Robiolo G and Oroscom R 2012 *An Alternative Method Employing Uses Cases for Early Effort Estimation* (Universidad Austral and Universidad Argentina de la Empresa).

[12] Madhavji H and Wilding M 2009 The Impact of Non-Technical Factors on Software Architecture (Canada: University of Western Ontario and IBM Canada Ltd).

Stakeholder Definition for Indonesian Integrated Agriculture Information System (IAIS)

Halim Budi Santoso[1]; Rosa Delima[2]

[1] Information System Department; Faculty of Information Technology; Duta Wacana Christian University

[2] Informatics Department; Faculty of Information Technology; Duta Wacana Christian University

Abstract. Stakeholders plays an important roles to determine the system requirements. Stakeholders are people or organizations that has an interest to the enterprise. Timely and effective consultation of relevant stakeholders is a paramount importance in the requirements engineering process.From the research and analysis of system stakeholder finds that there are four stakeholder groups in IAIS. Stakeholder analysis is being implemented by identifying stakeholder, stakeholder category, and analysis interaction between stakeholders.

1. Introduction

Indonesia is one of the biggest agricultural countries in the world. Wide areas in Indonesia are good places to grow various plant and trees. Based on World Bank, Indonesia has 1.905 km^2 which 12% of the area grows agriculture product and bring benefit to the communities, either farmers or farmers communities. Agriculture sector also supports 12% of Gross Domestic Product from 2010 – 2013. Productivity in Agriculture sector needs to be improve. Increasing productivity will increase agriculture sector player's welfare. One of the approaches to increase productivity is by implementing information and communication technology for agriculture sectors. Integrated Agriculture Information System is a computer based system which consists of some application in Agriculture. This research sets up a study group focussing in Indonesia Integrated Agriculture Information System (IAIS). IAIS is a system to fulfil stakeholder's requirement. Stakeholder play an important role in this system as information source and information recipient in this system. Thus, information requirements of stakeholders should be analysed and documented and then adequate Information Systems (IS) should be developed.

Stakeholders participate in the system requirement engineering. Stakeholders are related to each other and interact with each other (Sharp, et. al., 1999). Interactions between stakeholders including: (1) Exchanging information; (2) Exchanging products; (3) instructions; (4) providing supporting tasks. By analysing stakeholders can help to understand the requirements of system for each stakeholders.

As the Sharp, et. al. (1999) has been studied that every system engineering needs stakeholders. IAIS also needs stakeholder for its system. Thus, in this research try to achieve goal to analyse and define the stakeholders. IAIS involves interorganizational stakeholders and by analysing the needs of the stakeholders can be useful as a mechanism to expose conflict that might be appeared in the future and different expectations from different stakeholders.

2. Literature Review

2.1. Agriculture Information System

In the world, Agriculture Information System is an interesting and recent research topic. This topic is an interdisciplinary topics between agriculture science and information technology. There are some application that is developed to help the farmers to increase the productivity. In Iran, Hosseini and Rezaei (2013) reported a Management Information Systems for agriculture. Management Information Systems are tools to give a farmers to access a wide range of integrated data sets. Rolling (1998) reported that an agricultural information systems is generated, transformed, transferred, consolidated, received, and fed back in such a manner that these processes function synergistically to underpin knowledge utilization procedures. Supporting the Hosseini and Rezaei (2013), Babu, et. al. (1997) consider that Management Information System in the agriculture can help farmers to: (1) plan the effective resources to cultivate the land and its crop; (2) choose the best alternatives to produce and take part in the cultivation; (3) manageable day to day operations for processing the land.

Another systems in Agriculture is agriculture trade systems that facilitate the farmers to maintain the crop price. This system connects the farmers with business people and company that work in the agriculture sectors. One of the system in crop trading and pricing is Infotrade which implements in Uganda and Portable Agricultural Market Information Collections (PAMIC) which implements in China (Xu, 2013a). Xu (2013a) in his report discovers that to strengthen the agricultural information monitoring, portable agricultural market information collections (PAMIC). This system is integrated with GPS Satellite Positioning, 3G Information real – time transmission, and stylized information collection. This system will help the productivity since it uses mobile technology to develop the systems. Wen, etc (2007) reported that mobile technology is suitable for the farmers since farmers are easy to get access to this technology comparing with other technologies.

Another systems developed for Agriculture sector is China Agriculture Monitoring and Early Warning System (CAMES) (Xu, 2013b). This system is developed based on the consideration for world of food security. Besides, China is also well known as a country that is prone to natural disasters (Xu, 2013b). Based on that problems, China tried to develop CAMES which has an objective to create an integrated agriculture multi-market, multi-product, multi-area, and multi-function. This system was real-time monitoring and analysing all kinds of variables, dynamic monitoring, evaluating, simulating, prospecting, and early-warning systems for agriculture product and productions. CAMES also provides knowledge management systems that has function to query agricultural product forecasting.

2.2. Farmers Needs in Agriculture Information System

ICT plays an important role as a new perspective to develop the agriculture sector. However, farmers still have problems to get information that is easily to be understood. This information will help farmers to decide agricultural production improvement. Thus, farmers will have a better result in decision and generate greater profit. Besides, implementation of ICT will increase their competitiveness for greater market (Cecchini and Scott, 2003; Courtright, 2004).

These the following farmer's information requirements for ICT implementations (Milavanovic, 2014):

1. Information on Crops: Farmers need some information related with crops and plants, such as: (1) categories of seeded crops; (2) size of the land with specific crops; (3) time of dropping seeds; (4) time to harverst; (5) yields, etc. These information are needed to help farmers to plan their crop.
2. Information on Production Techniques is an information to help farmers in new production techniques. It helps farmers to improve its ability to increase their production skills.
3. Information on Production Equipment and Agricultural Inputs is an information to help farmers to know the tools for soil and plant processing. These information also are able to help farmers learn how to use the tools and farmers equipment.
4. Market Information is an information to support farmers in gaining the best prices for their products, information on market of various agricultural products should be created. This information facilitates farmers to access prices of various markets.

Farmers need ICT application to support its daily operation in agricultural production. These applications will play great role in operation management and help to decide based on real time data (Milavanovic, 2014). Henderson, et. al., (2006) reported that by using internet can help to improve communication and to create business opportunities for agricultural communities. Farmers, researchers, cooperatives, suppliers, and buyers can use the internet to exchange ideas and information and to govern business as well.

Leitgeb, F., et. al (2011) suggested an Innovation system based on farmer's experience. Leitgeb, F., et. al (2011) claimed that farmer's experiments and innovations are crucial to prepare agriculture for global change. Cuban farmers participate to help building up a socially based knowledge pool and leads to agricultural development. Thus, it helps to build community based systems that can help the farmers share their knowledge to others and help to improve others knowledge. The innovative capacity of individuals and groups is a powerful element of Agricultural Innovation System that can contribute to the system's resilience.

3. Research Methodology

In conducting this research, the following methodology is done:

3.1. Initial Study for Implementation of ICT in Agricultural Sector

Initial study was conducted in 2015. There are 77 participants from Yogyakarta participated in this study. From the data gathering through interview, questionnaire, and study group. The participants are chosen from the farmer communities in Yogyakarta.This research gained and discovered these information as follows: (1) General characteristics of farmers in Yogyakarta are : 75% of the farmers are 40 years old, Farmers graduated from Senior High School, have been working as a farmers for more than 10 years; (2) The use of ICT and Internet in Agricultural sector is still limited, example farmers use smartphone to look information about agriculture; (3) Farmer lacks of IT literacy, especially related with the use of smartphone, computer, and internet; (4) 66% of the farmers are eager to join and participate in ICT training; (5) ICT infrastructure in rural areas is available, either through telephone or wireless network; (6) Computer application that farmers needs are information about prices, farming tools and technique, product marketing and sales, land processing technique, and post-harvest product processing. This initial study is a very important foundations to support the next step in defining the stakeholders and its information (Delima and Purwadi, 2015).

3.2. Literature Review of Stakeholder Requirement in Agriculture Information Systems

After conducting the initial / preliminary study, the next phase is literature study related with stakeholder's information requirement (Milavanovic, 2015; Delima and Purwadi, 2015; Leitgeb, F., et. al., 2011). Purpose of this study to discover the information needs for each stakeholders involved in agriculture systems. From the literature study, it is found that farmers needs the following information:

1. Information related with crop which is needed to help farmers to plant the crop.
2. Information on tools and technique to cultivate the crop, including the land preparation, land selection, crop selection, harvest, and post-harvest processing.To have the latest and newest farming technique, Integrated Agricultural Information System should cooperate with researchers who have a passion and research in agriculture sectors.
3. Market information is needed for farmers as useful information's to broaden the market. Farmers also need to have access to business sector to gain information about crop price, seeds price, product price, tools and equipment price (Milavanovic, 2014)
4. Farmers has much experience. Delima and Purwadi (2015) reported that most of Indonesian farmers has been working for 10 years. Thus, it is a good perspective to develop IAIS by conducting participatory and knowledge sharing (Leitgeb, F., et. al., 2011). Participatory encourage farmers to be participate more actively and share with other farmers. Thus, it can broaden their knowledge.

5. Integrated Agriculture Information System also has to cooperate with government. Government can help development of IAIS by give some support in regulations and funds.

3.3. Set up the Stakeholders Category, Stakeholders Maps, and Stakeholders Matrix.

Stakeholder category, stakeholder maps, and stakeholder matrix are needed to draw stakeholders involved and participate in IAIS. By analyse and grouped into several groups based on its power and interest

4. Analysis and Discussions

4.1. Stakeholder Category

Based on literature study, there are 23 stakeholders involved and participate in IAIS. Those stakeholders can be categorized into 4 following categories:(1) Farmers; (2) Researchers; (3) Public Sector; (4) Business Sector. These categories to ease in fulfilling information needs for each stakeholder. Stakeholder category identifies stakeholder group and its member for each category. Table 1 below gives clear definition of stakeholder categories, its roles, and its member for each category.

Table 1. Stakeholder Category

Stakeholder Category	Roles	Members
Farmers	The main actor in the Integrated Agriculture Information System; Consumer of Financial Services; Information sources; Organize farmers and give accesses to third parties	Farmers; Farmers Union / Farmers Community
Researchers	Information sources; research and development; interaction between researchers and farmers to develop new technology; accompaniment and provide training to farmers about new technology and new development in agriculture sector.	Faculty Member of Agricultural Science; Faculty member of Information Technology; Higher Education Researchers
Public Sector	Provide information technology infrastructure, physical infrastructure (like reservoir, land); Distribute fertilizer, seeds, farming tools and equipment; Regulate price, standardization; Provide information and data related with Agriculture.	Ministry of Communication and Information; Ministry of Public Workers; Local Government; Ministry of Agriculture; Welfare NGO; Agricultural NGO
Business Sector	Financial services and insurance; Produce pesticide, fertilizer, farming tools and equipment; Distibute and selling crops; Information sources in agriculture sector	Transportation and Courier Services; Fertilizer and pesticide manufacture; Distributor; Suppliers for farm equipment's and tools;e-Commerce and e-Channels; Banking and Financial Institutions; Telecommunication providers / operator; Cooperation; Information Technology Practitioners; Retailers (including online store); General Distributors; Distributor of Agricultural Products; Distributor of Agricultural Derivatives Products; Supplier of Agricultural Equipment;

Table 1 shows the stakeholder group, roles, and its member (who is categorized in this part). The main stakeholder for this information system are farmers and farmer communities. They are involved because it helps to organize the farmer.

The second stakeholder in this IAIS are researchers from two major, Agriculture science and Information Technology. Agriculture Science Researchers are needed to give some input for the agriculture systems and technics, new technologies in agriculture, crop diseases and its medication, crop selection process and technics. Researchers from information technology is competent and capable to plan, develop, and maintain the information system and its architecture.

Public sector involves some ministries who are interested and have some power in this system. Government, either municipalities or ministry, involve in Integrated Agricultural Information System (IAIS) by providing access and infrastructure to the Information Technology. Government and NGO also can help to support funding to develop IAIS. Business sectors provide some support starting from financial services, pesticide, fertilizer, farming tools and equipment, distribution path and courier services.

4.2. Stakeholder Power and Interest

The next process is analysing power and interest of each stakeholder in relationship with IAIS. This stakeholder analysis is the technique used to identify who participates actively in IAIS. This research analyse the power and interest of each stakeholders based on its function in agriculture sector. The bigger the function in agriculture sector will make the stakeholder have bigger power and interest in Indonesia Integrated Agriculture Information System.

In table 2 below, there are 4 quadrant in this matrix, as follows: (1) Stakeholders which have high power and high interest in IAIS should fully engage and make the greatest efforts to satisfy. Stakeholders in the area of key players also can make the development of IAIS fail if study group is not maintain well; (2) Stakeholders which have high power and less interest should work not too much with these stakeholders because they may become bored with the report that system developers send; (3) Stakeholders which have low power and high interest should be informed adequately. IAIS developer also should talk to them to ensure that there is no major problem arising; (4) Stakeholders which have low power and low interest only monitor and keep them informed in development of IAIS.

Table 2. Stakeholder Power and Interest Matrix

Keep Satistifed (Low Interest, High Power)	**Key Players (High Interest, High Power)**
Ministry of Communication and Information Local Government Telecommunication Operators Ministry of Public Works Trasportation Services Ministry of Research and Higher Education (Medium Interest, Medium High Power) Banking (Medium Interest, Medium High Power) Cooperation (Medium Interest, Medium High Power)	Ministry of Agriculture Farmers Distributor of Agriculture product Farmers Union
Minimal Efforts (Low Interest, Low Power)	**Keep Informed (High Interest, Low Power)**
NGO in Public Welfare IT Academician IT Practitioners Distributor of Derivatives Product General Distributor (Medium low interest, medium low power)	NGO in Agriculture Manufacture for pesticide and fertilizer Supplier of farming tools and equipment e-Commerce (Medium high power, Medium low interest) Academician in agriculture sector Agriculture instructor / advisor

4.3. Stakeholder Interaction Matrix

Stakeholders in IAISneeds different kind of information. Each stakeholder with different role needs different information. This information comes from interaction between stakeholders. Thus, stakeholder interaction and information matrix should be set up to identify information needed for every stakeholders. There are two parts in stakeholder's interaction matrix. One stakeholder plays as an information source and another as an information recipient. This matrix can be found on table 3 below:

Table 3. Stakeholder Interaction and Information Needed

Source	**Recipient**	**Information**
Farmers	Researchers	Current state information related to technology, tools, equipment, and treatement in farming
	Public Sector	Information related with phase in farming; current planting and its schedule for every phase; Information related with current crop diseases, obstacles and problems in farming, regulation issues in agriculture; needs for farming.
	Business Sector	Information related with agriculture products (fertilizer, pesticide, tools, equipment); informtion related with financial aids and funding
Researchers	Farmers	New issue in agriculture; New method for planting and farming; Research finding in agriculture

Source	Recipient	Information
	Public Sector	Research finding in agriculture; community empowerment program for agriculture sector
	Business Sector	Research finding in agriculture; community empowerment program for agriculture sector
Public Sector	Farmers	Issues in agriculture policy and regulation; Information relates with planting calendar; Weather forecast; funding and grant from government and third parties; research and development in agriculture technology; networking in farming and agriculture
	Researchers	Research funding; policy in agriculture and farming
	Business Sector	Issues in agriculture policy and regulation; Issue in standardization; government planning in agriculture sector
Business Sector	Farmers	Information for prices (fertilizer, seeds, pesticide, tools and equipment); Information for funding and grant from private sector; information and technology in food processing, food packaging. Information for yield price.
	Researchers	Research funding and grant; Joining CSR program to empower the farmer communities.
	Public Sector	Information for yield price; information product quality in agriculture and farming; information product and materials in agriculture which private sector provides.

Table 3 reflects interaction between stakeholders specifically. This interaction happens because of the information exchange among the stakeholder. IAIS has to be able to link information needs for the stakeholders automatically through fulfilment of the data and information. System developers, in this case the organization has a role as IAIS active user.

5. Conclusions and Future Works

1. Stakeholder analysis is the process to identify individual, groups, organizations, or government that are affected by the systems. IAIS has 23 stakeholders involved from academician, farmers, government, NGO, and private sector / business sector. Those 23 stakeholders are groups into 4 stakeholders: (1) Farmers; (2) Researchers; (3) Public Sector; (4) Business Sector. Those stakeholders are grouped based on its interest and information needed in IAIS.
2. Development of IAIS also conducted stakeholder'srelationship and information sharing for among those stakeholders. While the information needs between stakeholder is able to shape the interaction among stakeholders. There are 12 interactions which provides different information between the information source and recipient.
3. Stakeholders are also analyse and grouped into several groups based on its power and interest. Those to reflect the participation of each stakeholders in the development of IAIS and interest in implementation of IAIS.

The future works to develop the IAIS are: (1) Developing Business Architecture as a foundation for business rules and business processes for IAIS; (2) Identifying the Data and Application Architecture based on Vision Architecture and Business Architecture; (3) Developing technology framework to implement and build IAIS.

6. References

[1] Babu R A, Y P S and K S R 1997 *Managing Human Resources within Extention.* (Rome: Food and Agriculture Organization of the United Nations).#

[2] Cecchini S and Scott C. 2003 Can Information and Communications Technology Application Contribute to Poverty Reduction *Lessons from Rural India* **10(2)** pp 73 - 84.

[3] Courtright C 2004 Which Lessons Are Learned? Best Practices and World Bank Rural Telecommunication Policy *Information Society* **15(2)** pp 345 - 56.

[4] Delima R and Purwadi J 2015 *Studi Awal Identifikasi Penerapan Teknologi Informasi di Bidang Pertanian Menggunakan Pendekatan Partisipatif* (Yogyakarta: Universitas Kristen Duta Wacana).

[5] Henderson J R, Akridge J T and Dooley F J 2006 Internet and e-Commerce Use by Agribusiness Firms *J. of Agribusiness* **24(1)** pp 17-39.

[6] Hosseini S M and Rezaei A 2013 Developing an Information System for Sustainable Natural Resource Management in Alborz Watershed, Northern Iran *System Practise and Action*

Research **26** pp 131 - 152.
[7] Leitgeb F, Funes-Monzote F R, Kummer S and Vogl C R 2011 Contribution of farmers' experiments and Innovations to Cuba's Agricultural Innovation Systems *Renewable Agriculture and Food Systems* **26(4)** pp 354 - 67.
[8] Milavanovic S 2014 The Role and Potential of Information Technology in Agricultural Improvement *Economics of Agriculture* **61(2)** pp 471 - 85.
[9] Rolling N 1998 *Extension Science: Information Systems in Agricultural Development.* (Cambridge: Cambridge University Press).
[10] Sharp H, Finkelstein A and Galal G 1999 Stakeholder Identification in the Requirements Engineering Process *Database and Expert Systems Appl.* (Florence: IEEE Computer Society Press) pp. 387 - 91.
[11] Wen G, Zetian F, Daoliang L, Longyong Y, Jian Z and Xiaoshuan Z 2007 AgriInfo: an Agricultural Information System Based on a Call Center in China *New Zealand J. of Agricultural Research* **50** 797 - 806.
[12] Xu S 2013 *Enhancing Agricultural Market Information : China's Experience* (Geneva: UNCTAD Global Commodities Forum).
[13] Xu S W 2013b Construction of China Agriculture Monitoring and Early-Warning Systems. *World Agricultural Outlook Conf.* (Berlin: Springer) pp 1-14.

Analysis of usability factors affecting purchase intention in online e-commerce sites

R A Perdana and A Suzianti

Industrial Engineering Department, Universitas Indonesia, Depok, Indonesia

E-mail: raihan.ap23@gmail.com

Abstract. The growing number of internet users plays a significant role in the emergence of a variety of online e-commerce sites to meet the needs of Indonesians. However, there are still some problems faced by the users in using e-commerce sites. Therefore, a research related to user experience on their purchase intention to foster e-commerce sites is required. This study is conducted to find out the relationship between usability factors on e-commerce users' purchase intention using a case study by using SEM to analyse the usability of the website. The result of this study shows that credibility, readability and telepresence are usability factors that directly affect purchase intention, while simplicity, consistency and interactivity are usability factors that indirectly affect purchase intention. Therefore, we can conclude that Indonesian consumers are on the Early Majority phase in adopting Company A.

1. Introduction

With around 255.5 million citizens, Indonesia is the fourth most populated country in the world. Population growth in Indonesia is also followed by the continuous development of technology, thus increasing the number of internet usage in Indonesia for about 3% every year [8].

Online shopping is one of the use of internet technology. With the provision of convenience, the number of consumers who shop online increases up to 7,4 million by 2015. The amount of online transaction in online E-commerce sites also increases. Total transaction recorded in 2015 was US$ 3.56 billion and it is predicted to reach US$ 4.96 billion in 2016 due to the increasing affordability of upper middle class. According to a preliminary study conducted using online questionnaire on 123 respondents, it is known that 84.6% of them have experienced shopping online on e-commerce sites. This shows that online shopping is starting to be common practice for Indonesian citizens.

The impacts of website design on customers' purchase within the e-commerce scope could not be fully understood without doing website usability evaluation (DeLone and McLean, 2003) in [4]. Developing a useful website is very important on the success of e-business, because online consumers are only able to touch, feel, search and see the products (and service) through website. Developing a useful website is also known to give positive impact on online shop, increase loyalty and returning visits, and stimulate online purchase [4]. Some scientists have also done a research on the relationship between usability construct and various variables, such as between website usability and users' perception, action, and determinants of user perceptions.

The measurement discussed in this study is about UX, focusing on analyzing usability factors affecting purchase intention in Indonesia's online e-commerce sites. E-commerce sites will be represented using Company A. The purposes of this study are: 1. Knowing the usability factors affecting Company A users' purchase intention; 2. Knowing the relationship between usability factors and

Company A users' purchase intention; 3. Giving recommendation to Company A in developing their facility according to User Experience. The result of this study is expected to help in developing or designing e-commerce website that could satisfy Indonesian users' needs.

2. Preliminary concepts

2.1. User experience (UX)

UX quantitative model concept constructed by [6] is the most comprehensive one. The model is consisted of overall UX and three UX elements, namely usability, affect and user value (see figure 1). UX of an operating system is also affected by user interface, therefore a sufficient interactive system is needed. Interactive system provides a broad and complex functionality, and UX quality highly depends on its usability aspect.

2.2. Usability

According to ISO 1998, usability refers to the extent to which a product could be used by specific users to reach specific purposes with effectiveness, efficiency, and satisfaction in specific usage context. Developing a useful website is an important aspect in the success of e-business, because the internet is rapidly developing. Website usability refers to the extent to which a website could be used by the users to reach specific purposes in accessing with effectiveness, efficiency, and satisfaction in specific usage context.

Website usability construct is built up by Simplicity, Readability, Consistency, Interactivity, Learnability, Navigability, Content Relevance, Supportability, Credibility, and Telepresence factors [4]. A previous research also saw the relationship between website usability and purchase intention, thus forming a usability model as seen in figure 2 below.

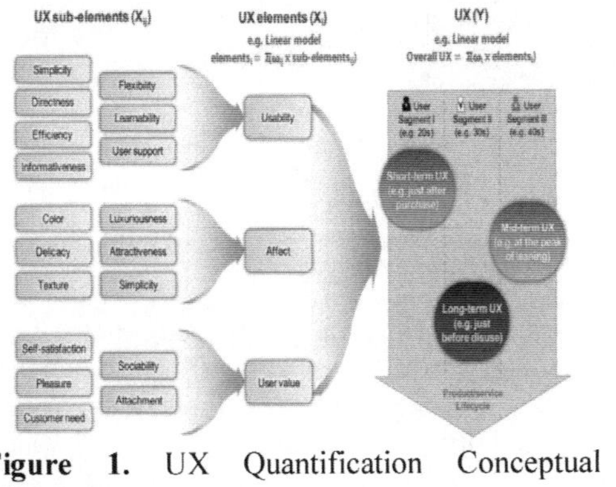

Figure 1. UX Quantification Conceptual Diagram

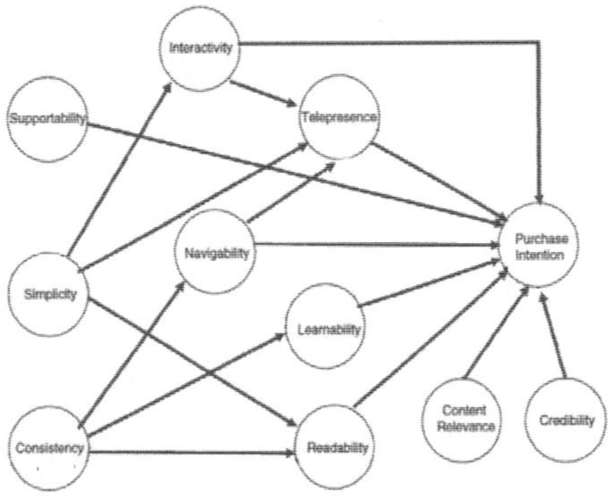

Figure 2. Framework Website Usability Model

2.3. Technology adoption lifecycle (TALC)

Technology Adoption Lifecycle (TALC) is a socio-economic model which explains how a new idea and technology is distributed through consumers' different cultures. This model describes the acceptance of a new product or innovation according to demographic and psychological characteristics of the recipient group. In other words, technology adoption lifecycle describes how a market develop new products. This TALC model elaborates the adoption stages of a new product or innovation, which is divided into 5 categories, namely innovators, early adopters, early majority, late majority, and laggards. They are illustrated by a belt-shaped curve [5].

2.4. Theoretical Framework

In general, the theoretical framework of this paper is presented in figure 3. This study focuses on the relationship between usability factors that affects e-commerce users' purchase intention. The analysis is done using Structural Equation Modelling (SEM). Company A is a representation of the market condition in Indonesia. The results will later be elaborated and analyzed further using Technology Adoption Lifecycle (TALC). The final results of this study are expected to provide improvement strategies that could be implemented for the development of Company A.

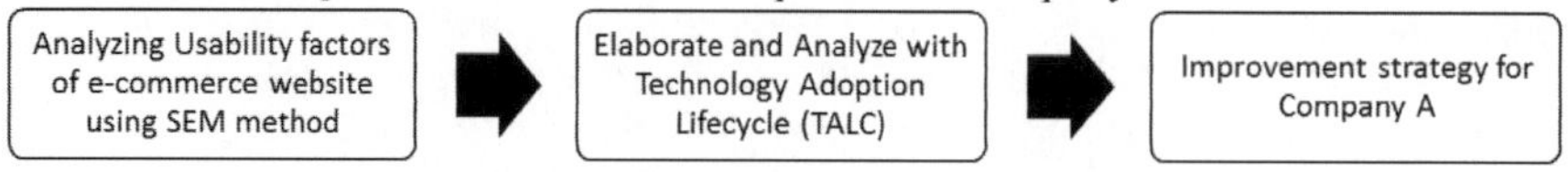

Figure 3. Theoretical Framework of this Study

3. Method of research and results

3.1. Method of research

The data type used in this study is primary data. The data is collected by distributing research instrument in a form of quantitative questionnaire. Google Forms service is used to make the questionnaire and it is then distributed online. The questionnaire uses likert scale of 5 to find out the level of agreement. Target respondents of this study are Company A users residing in DKI Jakarta and West Java.

Table 1. Research hypothesis

Hypothesis	Information
H1	*Simplicity* will positively affect the *Interactivity*
H2	*Simplicity* will positively affect the Telepresence
H3	*Interactivity* will positively affect the *Telepresence*
H4	*Navigability* will positively affect the *Telepresence*
H5	*Simplicity* will positively affect the *Readability*
H6	*Consistency* will positively affect the *Readability*
H7	*Consistency* will positively affect the *Navigability*
H8	*Consistency* will positively affect the *Learnability*
H9	*Supportability* will positively affect the *Purchase Intention*
H10	*Credibility* will positively affect the *Purchase Intention*
H11	*Content Relevance* will positively affect the *Purchase Intention*
H12	*Interactivity* will positively affect the *Purchase Intention*
H13	*Telepresence* will positively affect the *Purchase Intention*
H14	*Readability* will positively affect the *Purchase Intention*
H15	*Navigability* will positively affect the *Purchase Intention*
H16	*Learnability* will positively affect the *Purchase Intention*

This study uses usability construct model on purchase intention. This model constructed by [4] could provide a broader information and knowledge because it takes usability construct, which directly and indirectly related to purchase intention, into account. There are 11 latent variables in this study, namely Simplicity, Readability, Consistency, Interactivity, Learnability, Navigability, Content Relevance, Supportability, Credibility, Telepresence, and Purchase Intention. This study model is resulting in 16 hypotheses to be tested as shown in Table 1.

Validity and Reliability testing was conducted to determine the feasibility of the data obtained before inputting data to LISREL 8.80 software. SEM method used to analyze the relationship between factors of usability. Further analysis is then done using TALC method to determine Company A's users' characteristics so a solution for improving Company A's website usability could be given.

3.2. Data processing result

This study is conducted using SEM method (LISREL 8.80 software). By Confirmatory Factor Analysis (CFA) in the measurement model, the first step is model specification. There are five exogenous and six endogenous variables with 3 to 5 observed variables each as seen in Appendix A. The second step is identification which needs an over-identified model. The model in this study is already categorized as over-identified with 875 degrees of freedom score. The third phase is estimation using Maximum Likelihood estimator which needs a minimum data of 220 respondents, and the data collected is of 729 respondents, thus making it sufficient. In goodness of fit test, validity test and reliability test are conducted. Validity tests shows minimal t-value as 1.96 and standardized loading factor (λ) > 0.50, while reliability test shows construct reliability value as (CR) $\geq$ 0.70 and variance extracted (VE) $\geq$ 0.50. From the calculation, we could see that there are some values that are below standard, therefore a respecification is conducted to delete S3, S4, I1, N2 variables (marked by * in Appendix A). Then, goodness of fit test is reconducted and the result shows that the model is valid and reliable with 11 latent variables and 40 items used. Afterwards, structural model goodness of fit test is conducted, and the result shows that the entire models are good fit (Table 2). Structural Model Analysis is also seen from t-value path hybrid model to know the relationship between latent variables from study model.

Table 2. Structural Model Fit

Goodness of Fit	Recommended value	Result	Information
Chi-square (χ2)	Small value	2827.75	Bad Fit
p-value	p>0.05	0.00	
Goodness of Fit Index (GFI)	GFI≥0.90	0.84	Marginal Fit
Root Mean Square (RMR)	RMR≥ 0.05	0.05	Good Fit
Root Mean Square Error of Approximation (RMSEA)	RMSEA ≤ 0,08	0.061	Good Fit
Non-Normed Fit Index (NNFI)	NNFI≥0.90	0.98	Good Fit
Normed Fit Index (NFI)	NFI≥0.90	0.97	Good Fit
Adjusted Goodness of Fit Index (AGFI)	AGFI≥0.90	0.81	Marginal Fit
Relative Fit Index (RFI)	RFI≥0.90	0.97	Good Fit
Incremental Fit Index (IFI)	IFI≥0.90	0.98	Good Fit
Comparative Fit Index (CFI)	CFI≥0.90	0.98	Good Fit
CN	CN≥200	234.1	Good Fit

4. Analysis

4.1. T-value analysis

T-value could describe the influence between one latent variable on another. Reference [10] explains that a variable is said to have an influence if the t-value is ≥ 1.96. According to structural model t-value, there are 10 hypotheses Accepted or above standard and 6 hypotheses Rejected or below standard. T-value of H1, H2, H3, H5, H6, H7, H8, H10, H13, and H14 are above 1.96. And t-value of H4, H9, H11, H12, H15, H16 are below 1.96. In other words, the result of this study shows that Company A users significantly feel that readability, telepresence and credibility could affect users' purchase intention in online e-commerce sites. However, supportability, interactivity, navigability, learnability, and content relevance factors of Company A have not significantly affect purchase intention. Moreover, Company A's simplicity and consistency through readability factor and interactivity through telepresence factor also indirectly affect users' purchase intention. In other words, Company A consumers are now paying more attention to non-technical aspect such as transaction safety and confidentiality, easiness in understanding the website, as well as sense of belonging in Company A when purchasing.

4.2. Technology adoption lifecycle (TALC) analysis

According to Technology Adoption Lifecycle, which has five adoption stages: Innovators, Early Adopter, Early Majority, Late Majority and Laggards, we can conclude from the result of this study that Company A's consumers are in Early Majority stage in accepting Company A as one of Indonesia's online e-commerce website. In this phase, consumers are considered pragmatic because they tend to be cautious towards new product or innovation [5]. Consumers tend to be influenced by personal information source from the ones around then and will gather as much information relating to the advantages of the available product or innovation. Furthermore, consumers also depend on the other consumers' experience in using a product or innovation. Consumers in this phase tend to avoid the risk of using new product or innovation, so they have relatively long time in making decisions compared to the consumers in Innovators and Early Adopters stage. With Early Majority stage market condition, companies have a few choices in developing marketing strategy, namely:

1. As a marketplace, Company A could do a collaboration and provide education to SME community about the importance of online media in reaching broader market, so that SME business operators could be interested in marketing their products through Company A.
2. Company A could design a website by combining pictures and font with a minimum standard size of 14 pixels [2]. This font size change will make it easier for users to read and understand the information stated on the website.
3. The company should provide education and make promotions regarding to shopping security and provide warranty for every purchase made in Company A. Furthermore, to increase Company A's credibility on Indonesian consumers' eyes, those efforts should be compensated by a good customer service support.

5. Conclusion

This study is conducted to identify the impact and relationship between usability factors on purchase intention in Company A. According to the model studied, usability factors that directly give positive impact on purchase intention are readability, telepresence, and credibility. Usability factors that indirectly give positive influence are simplicity, consistency, and interactivity.

The result of this study shows that according to Technology Adoption Lifecycle, Company A's consumers are now on Early Majority stage in adopting Company A as Indonesian online e-commerce website. Consumers in this stage will tend to be more cautious in accepting innovation that they are considered pragmatic. Consumers will gather as much information as possible in accepting new information to reassure that the innovation is stable and that its benefit has been proven. In this case, Indonesian consumers are still more concerned with credibility, sense of belonging, and readability factor of an online e-commerce website.

Acknowledgement

We would like to give our sincerest gratitude to Direktorat Riset & Pengabdian Masyarakat (DRPM), Universitas Indonesia, for funding this study. Thus, this study can be properly carried out and produce such results. This research was also made possible by the support and assistance of several people. We would like to thank for their valuable and constructive opinions.

Appendix A. Definition and instrument items of factor used

Construct	Definition	Instrument Items	Code	Ref
Simplicity (S)	Provision of minimum contents and functions within a website	The structure of the website is succinct	S1	Lee and Kozar (2012)
		I can comprehend most components of a page within seconds	S2	
		*The website has components that are not necessary	S3	
		*There are redundant components in the website	S4	
Readability (R)	Extent to which website components are well organized and easy to read and understand	The website's wording is clear and easy to understand	R1	Lee and Kozar (2012)
		The website has enough white space (or margins) to make it readable	R2	
		Every page contains the appropriate amount of components to fit into a page	R3	
		The website uses colors and structures that are easy on the eyes	R4	
Consistency (Cs)	Consistent location of page components within and across pages	The website repeats the same structure, components, and overall look across pages	Cs1	Lee and Kozar (2012)
		The website contains similar components across web pages	Cs2	
		Web pages in the website are consistently designed	Cs3	
		Each web page on the website is of similar design	Cs4	
Interactivity (I)	Website's ability to create vivid interaction and communication with users	*The website provides an appropriate amount of interactive features (e.g., graphics, pop-up windows, animation, music, voices)	I1	Lee and Kozar (2012)
		The website contains components to help the interaction between it and consumers	I2	
		Interactive features of the website are vivid and evoke responses	I3	
		The website provides features for interactive communication between consumers, or between consumers and the online company	I4	

Appendix A. Definition and instrument items of factor used (cont.)

Construct	Definition	Instrument Items	Code	Ref
Learnability (L)	Easy to learn the main functionality and gain proficiency to complete the tasks	The contents provided by the website are easily understood	L1	Lee and Kozar (2012)
		The website is designed for easy understanding	L2	
		I can easily remember how to reach the same page when I visit next time	L3	
		As time passes, I am more accustomed to the website with less effort	L4	
Navigability (N)	Capability to provide alternative interaction and navigating techniques	The website provides multiple search features (e.g., search engine, menu bar, go-back-and-forward button, etc.) to obtain the target information	N1	Lee and Kozar (2012)
		*The web page that I am looking for can be reached through multiple pathways	N2	
		There are multiple ways to access the web page that I am looking for and/or return to shopping menus	N3	
		It is very easy to locate what is needed in this website	N4	
		The website keeps the user oriented as they shop	N5	
Content Relevance (CR)	Extent to which the content is up-to-date and pertinent	The website contains in-depth information	CR1	Lee and Kozar (2012)
		The website provides up-to-date information	CR2	
		The scope of information provided by the website is appropriate	CR3	
		The information provided by the website is accurate	CR4	
Supportability (Sp)	Additional information and support mechanisms readily available to enhance the website use	While visiting the website, I feel that I can get just-in-time support anytime I need it	Sp1	Lee and Kozar (2012)
		The website provides features to ask for help anytime I need	Sp2	
		Getting support through a series of options is easy and convenient	Sp3	
Credibility (C)	A holistic concept that covers an online user's perception of security, privacy, and reliability during the navigation	I feel safe in my transactions with the website	C1	Lee and Kozar (2012)
		I trust the website to keep my personal information safe	C2	
		I trust the website administrators will not misuse my personal information	C3	
		The website is stable to use	C4	
		The website provides detailed information about security features	C5	
Telepresence (T)	Sense of presence in a virtual environment created by a computer/ communication medium	I felt empathy with the website	T1	Lee and Kozar (2012)
		I feel I have personal ties to the website	T2	
		I feel as though I am emotionally connected to the website	T3	
		I feel as though I am taking part with the website	T4	
Purchase Intention (PI)	How likely the user wishes to buy products from the website	I plan to shop at Company A's website in the future	PI1	Wang, Yeh, and Liao (2013)
		I intend to shop at Company A's website in the future	PI2	
		I predict I would buy a product from Company A's website in the future	PI3	

References

[1] Baleghi-Zadeh S, Ayub A F, Mahmud R, and Daud S M 2014 Learning Management System Utilization among Malaysian Higher Education Students: A Confirmatory Factor Analysis *Education & Human Development* **3** 369–386

[2] Banerjee J, Majumdar D, Pal M S and Majumdar D 2011 Readability, subjective preference and mental workload studies on young indian adults for selection of optimum font type and size during onscreen reading *Al Ameen Journal of Medical Sciences* **4** 131–143

[3] George D and Mallery M 2010 SPSS for Windows Step by Step: A Simple Guide and Reference (Boston: Pearson)

[4] Lee Y and Kozar K A 2012 Understanding of website usability: Specifying and measuring constructs and their relationship *Decision Support Systems* **52** 450–463

[5] Moore G A 2009 *Crosing the chasm: Marketing and selling technology project* (United States of America: HarperCollins Publisher, Inc.)

[6] Park J, Han S H, Kim H H, Oh S and Moon H 2013 Modelling user experience: A case study on a mobile device *International Journal of Industrial Ergonomics* **43** 187–196

[7] Pucillo F and Cascini G 2013 A framework for user experience, needs and affordances *Design Studies* **35** 160–179

[8] Singapore Post Ecommerce 2014 Indonesia's ecommerce landscape 2014: Insights into one of asia pacific's fastest growing markets (Singapore: Singapore Post Ltd)

[9] Wang YS, Yeh CH and Liao YW What drives purchase intention in the context of online content services? The moderating role of ethical self-efficacy for online piracy *International Journal of Information Management* **33** 199–208

[10] Wijanto S H 2008 Structural equation modelling dengan LISREL 8.8 (Yogyakarta: Graha Ilmu)

IOP Conf. Series: Materials Science and Engineering **185** (2017) 012016 doi:10.1088/1757-899X/185/1/012016

Cluster Analysis of Indonesian Province Based on Household Primary Cooking Fuel Using K-Means

S N Huda[1]

[1]Dept. of Informatics, Universitas Islam Indonesia, Yogyakarta, Indonesia
E-mail: [1]sheila@uii.ac.id

Abstract. Each household definitely provides installations for cooking. Kerosene, which is refined from petroleum products once dominated types of primary fuel for cooking in Indonesia, whereas kerosene has an expensive cost and small efficiency. Other household use LPG as their primary cooking fuel. However, LPG supply is also limited. In addition, with a very diverse environments and cultures in Indonesia led to diversity of the installation type of cooking, such as wood-burning stove brazier. The government is also promoting alternative fuels, such as charcoal briquettes, and fuel from biomass. The use of other fuels is part of the diversification of energy that is expected to reduce community dependence on petroleum-based fuels. The use of various fuels in cooking that vary from one region to another reflects the distribution of fuel basic use by household. By knowing the characteristics of each province, the government can take appropriate policies to each province according each character. Therefore, it would be very good if there exist a cluster analysis of all provinces in Indonesia based on the type of primary cooking fuel in household. Cluster analysis is done using K-Means method with K ranging from 2-5. Cluster results are validated using Silhouette Coefficient (SC). The results show that the highest SC achieved from K = 2 with SC value 0.39135818388151. Two clusters reflect provinces in Indonesia, one is a cluster of more traditional provinces and the other is a cluster of more modern provinces. The cluster results are then shown in a map using Google Map API.

1. Introduction

Human can not live without energy, in every activity from transportation, work, to even the most basic and needed activity, eating (cooking), we need energy. Each household definitely provides installations for cooking. Kerosene, which is refined from petroleum products once dominated types of primary fuel for cooking in Indonesia. Whereas the processing of crude oil into kerosene has an expensive cost and small efficiency, in addition to the product can be processed further to produce Avtur (Aviation Turbine) whose efficiency is higher. Indonesian government had rolled out a program to replace kerosene which still had high consumption rate, into LPG (Liquid Petroleum Gas) for cooking.

LPG does have better efficiency compared to kerosene, LPG consumption has a calorific value of 11254.61 Kcal / Kg compared to kerosene amounted to 10478.95 Kcal / Kg, with one liter of Kerosene LPG is equivalent to 0.57 Kg LPG. In addition, carbon emissions from LPG is smaller than the carbon emissions from kerosene. However, LPG supply is also limited. This can be seen from the results of LPG production, in 2000 the production of LPG in Indonesia still met domestic needs, so it did not need to be imported, whereas in 2010 Indonesia needed to import up to 1 million tons of LPG [1]. In the other hand, the utilization of natural gas by pipeline to the household does not require the cost for packaging tubes, but the installation of gas lines currently only covers a small part of settlements in several cities.

In addition, with a very diverse environments and cultures in Indonesia led to diversity of the installation type of cooking. We are still relatively easy to find a wood-burning stove brazier in Central Java. But if we make a survey in Jakarta, stove brazier is hard to be found.

The government is also promoting alternative fuels, such as charcoal briquettes, and fuel from biomass [2]. The use of other fuels is part of the diversification of energy that is expected to reduce community dependence on petroleum-based fuels.

The use of various fuels in cooking that vary from one region to another reflects the distribution of fuel basic use by household. There are some provinces in Indonesia which rely heavily on the use of LPG as the primary cooking fuel. Several other provinces use wood as the dominant primary fuel to cook. By knowing the characteristics of each province, the government can take appropriate policies to each province according each character. Therefore, it would be very good if there exist a cluster analysis of all provinces in Indonesia based on the type of primary cooking fuel in household.

For each province, the main fuel for cooking are calculated from the percentage of family using electricity, LPG, kerosene, charcoal / briquettes, wood, and other fuels. Other fuels here include households who do not cook.

Cluster analysis is done using K-Means method, where K-Means is well known for its simplicity. K-Means method has been used in many cluster analysis, such as in [3], k-means method is used to cluster soil character. In the same field, soil data clustering was also performed by [4] used two different methods, namely K-Means and Fuzzy C-Means uses soil data from Montenegro. The results of the K-Means soil clustering is visually displayed using the Google Map API.

In this paper, K-Means method is used to cluster Indonesian province based on the primary cooking fuel in household. Silhouette Coefficient is used to validate the result.

2. Literature Review

2.1. K-Means Clustering

Cluster analysis is a technique in finding the group in the data set in order for the data residing on one group has a close resemblance, and have a clear distinction with the other groups [5].

K-Means is a basic algorithm that is widely used for the data clustering [6]. Good clustering method will produce high quality clusters with high intra-class similarity (cohesive in one cluster) and low inter-class similarity (a clear difference between groups). The quality of a clustering method depends on the similarity measurement used in the method, its implementation, and the ability to find some hidden patterns. Cluster found using the K-Means algorithm has a convex shape and each cluster is represented by a center / centroid [3].

K-Means is a method for partitioning a set of data to treat the data as an object that has a location and distance to other data. K-Means will partition the object into k classes which are mutually exclusive, so that every object in each cluster is as close as possible to one another, and as far as possible with objects in other clusters. Each cluster is characterized by centroid, or midpoint. K-Means clustering algorithm is one of the most widely used, was first described by [7]. K-Means designed to cluster numerical data where each cluster has a central point, which is called the average. K-Means clustering algorithm is classified as partition or nonhierarchical method [8].

In this algorithm, the number of clusters k is assumed to be static. Then each data is allocated into the nearest cluster. After all data is allocated into k cluster, recalculate the midpoint of each cluster and reallocate each point to the cluster closest to the new midpoint repeatedly until the midpoint of each cluster does not have significant differences or any membership of each data is not changing [9].

To calculate the distance, we use Euclidean Distance as follows.

$$d_{ik} = \sqrt{\sum_{j=1}^{m}(x_{ij} - c_{kj})^2} \quad (1)$$

where:
d_{ik} = distance of i-th data to center of cluster k
m = number of attribute
x_i = i-th data
c_k = center of cluster k

2.2. Silhouette Coefficient
Silhouette coefficient is a method of interpretation for the cluster validation on data objects. This technique provides a brief graphical representation of how well each object is located in its cluster. Silhouette coefficient was first developed by [10].

A silhouette coefficient value of an object, egAj, is in the range between -1 to 1. The closer the object silhouette Aj to 1, the higher the degree of ownership of the object Aj in the cluster. While the value close to -1 (negative value) indicates that an object is likely to be on the wrong cluster, because the object similarity with other objects in the same cluster is low.

Silhouette coefficient test is a popular method that combines in a cluster cohesion and separation between clusters. For each data point, the calculation involves three steps.

1. For each object i in the cluster, calculate the average distance from the object to the entire objects in its own cluster. Call the average distance between the object i to all members of the cluster as ai.
2. For each object i and other cluster that is not cluster where i was in, calculate the average distance from the object to the entire objects in other cluster closest to cluster where i was (cluster neighbors). Call average distance to all objects in the neighboring clusters as bi.
3. For each object i, the silhouette coefficient obtained from:

$$(b_i - a_i)/\max(a_i, b_i) \quad (2)$$

The value of silhouette coefficient resides between -1 and 1. The negative value is not desirable because these values correlated with the case where the ai, as the value of the average distance to other point in the cluster, larger than the value of bi, the smallest average value to points on the other cluster. Silhouette coefficient value is expected to be positive (ai<bi) and ai is expected to be as close as possible to 0, where the silhouette coefficient reaches its maximum value 1 when ai = 0.

The average value of the Silhouette coefficient of the entire cluster will reflect the quality of the clusters. Based on [5] the average value of the Silhouette Coefficient (SC) is interpreted as a strong or weak classification as shown in Table 1.

Table 1. SC Interpretation

SC Value	Interpretation
0.71-1.00	Strong Classification
0.51-0.70	Good Classification
0.26-0.50	Weak Classification
0-0.25	Bad Classification

3. Dataset
The dataset used is the primary cooking fuel percentage of households by province in Indonesia in 2014 which are available from BPS as in Table 2. In Indonesia, we use some type of cooking fuel, such as electricity, LPG, kerosene, charcoal / briquettes, wood, and other fuels.

The clustering system is build using PHP. The dataset has to be normalized first. The data normalization is based on:

$$N_{ij} = (\frac{x_{ij} - \min(x_{1j,\ldots,}x_{nj})}{\max(x_{1j,\ldots,}x_{nj}) - \min(x_{1j,\ldots,}x_{nj})} \quad (3)$$

where

N_{ij} = Normalized value of i-th data in parameter j.

x_{ij} = value of i-th data in parameter j.

K on K-Means will be calculated from K = 2 to K = 5, then the silhouette coefficient test will be conducted. The system will display the most optimal silhouette coefficient from each K. Clustering output will be presented in the form of a cluster map using Google Map API.

Each cluster will be given a different marker, and each marker will be placed in the middle of their respective provinces. Province coordinate data refers to data from Google Map API. As for the representation marker will use a custom marker.

Table 2. Dataset used in experiment

Province	Percentage of Family Based On Province and Primary Cooking Fuel (2014)					
	Electricity	LPG	Kerosene	Charcoal/ Briquettes	Wood	Other
ACEH	0.11	64.61	3.97	0	28.82	2.49
SUMATERA UTARA	0.59	67.92	6.32	0.01	23.8	1.37
SUMATERA BARAT	0.31	23.47	25.54	0	49.22	1.46
RIAU	0.58	69.54	9.7	3.62	14.56	2
JAMBI	0.41	54.67	8.52	2.89	32.66	0.85
SUMATERA SELATAN	0.57	75.07	1.59	0.87	21.6	0.3
BENGKULU	0.37	62.59	1.77	0	34.09	1.19
LAMPUNG	0.27	53.39	0.43	0.03	45.25	0.62
KEP. BANGKA BELITUNG	0.54	56.64	27.53	0.03	14.76	0.49
KEP. RIAU	0.84	67.54	20.13	0.48	5.45	5.57
DKI JAKARTA	0.95	86.76	2.59	0	0.15	9.55
JAWA BARAT	1.06	76.34	0.46	0.02	19.59	2.53
JAWA TENGAH	0.5	63.45	0.25	0.07	33.47	2.25
DI YOGYAKARTA	0.4	54.19	0.57	0.37	33.93	10.54
JAWA TIMUR	0.5	62.77	0.81	0.03	33.56	2.33
BANTEN	0.79	76.75	0.54	0	19.79	2.12
BALI	0.36	62.62	0.98	0.02	28.85	7.18
NUSA TENGGARA BARAT	0.4	36.68	15.9	0	45.43	1.59
NUSA TENGGARA TIMUR	0.77	0.58	18.66	0	79.6	0.39
KALIMANTAN BARAT	0.63	68.3	1.19	0.07	29.42	0.4
KALIMANTAN TENGAH	0.36	16.65	39.41	0.05	42.72	0.82
KALIMANTAN SELATAN	0.16	39.94	24.34	0.03	34.11	1.42
KALIMANTAN TIMUR	0.58	79.94	8.8	0.1	9.04	1.55
SULAWESI UTARA	1.69	40.92	14.66	0.1	40.52	2.1

SULAWESI TENGAH	0.17	7.2	23.8	5.1	62.83	0.91
SULAWESI SELATAN	0.58	71.26	0.57	0.82	26.29	0.49
SULAWESI TENGGARA	0.22	16.28	25.18	2.36	55.18	0.79
GORONTALO	0.24	44.73	9.63	0.03	43.96	1.41
SULAWESI BARAT	0.33	42.72	1.27	1.41	53.92	0.36
MALUKU	0.1	0.48	48.23	0	50.61	0.58

4. Results

We build web based system to cluster provinces in Indonesia based on primary cooking fuel. It will iteratively calculate each cluster member using K = 2 to K = 5 and conduct the Silhouette Coefficient test from each K. The system will only display clustering result from the optimal Silhouette Coefficient result.From the experiment, the system generates the Silhouette Coefficient output as in Table 3.

Table 3. SC Result

K	SC Value
2	0.39135818388151
3	0.34231572024075
4	0.30726059280905
5	0.26847238402292

The optimal Silhouette Coefficient value is resulted from K = 2. The next optimal Silhouette Coefficient value is from K = 3. Therefore, the system will display a cluster membership table and map.

In K = 2, Cluster 1 consists of 22 provinces with characteristic of medium-high use of electricity, medium-high use of LPG, very low use of kerosene, very low use of charcoal / briquettes, very low-medium use of wood, and low-medium use of other fuel. This cluster depicts more modern provinces in Indonesia. Provinces in cluster 1 are Aceh, North Sumatera, Riau, Jambi, South Sumatera, Bengkulu, Lampung, Kep. Bangka Belitung, Kep. Riau, DKI. Jakarta, West Java, Central Java, DI. Yogyakarta, East Java, Banten, Bali, West Kalimantan, East Kalimantan, North Sulawesi, South Sulawesi, Gorontalo, and West Sulawesi.

While Cluster 2 consists of 11 provinces with characteristic of low use of electricity, low use of LPG, medium-high use of kerosene, low-medium use of charcoal/briquettes, high use of wood, and very low use of other fuel. This cluster depicts more traditional provinces in Indonesia. Provinces in cluster 2 are West Sumatera, NTB, NTT, Central Kalimantan, South Kalimantan, Central Sulawesi, South East Sulawesi, Maluku, North Maluku, West Papua, and Papua.

In K = 3, Cluster 1 consists of 5 provinces with characteristic of medium-high use of electricity, high use of LPG, very low use of kerosene, very low use of charcoal/briquettes, low use of wood, and high use of other fuel (other fuel include not cooking). This cluster depicts modern province in Indonesia. Provinces in cluster 1 are Kep. Riau, DKI. Jakarta, DI. Yogyakarta, Bali, and North Sulawesi.

Cluster 2 consists of 17 provinces with characteristic of low use of electricity, high use of LPG, very low use of kerosene, very low use of charcoal/briquettes, low-medium use of wood, very low use of other fuel. Provinces in cluster 2 are Aceh, North Sumatera, Riau, Jambi, South Sumatera, Bengkulu, Lampung, Kep. Bangka Belitung, West Java, Central Java, East Java, Banten, West Kalimantan, East Kalimantan, South Sulawesi, Gorontalo, and West Sulawesi.

Cluster 3 consists of 11 provinces with characteristic of very low use of electricity, very low use of LPG, medium-high use of kerosene, very low use of charcoal/briquettes, high use of wood, very low use of other fuel. This cluster depicts more traditional province in Indonesia. Provinces in cluster 3 are West Sumatera, NTB, NTT, Central Kalimantan, South Kalimantan, Central Sulawesi, South East Sulawesi, Maluku, North Maluku, West Papua, and Papua.

Figure 1 shows map output from the system using K = 2.

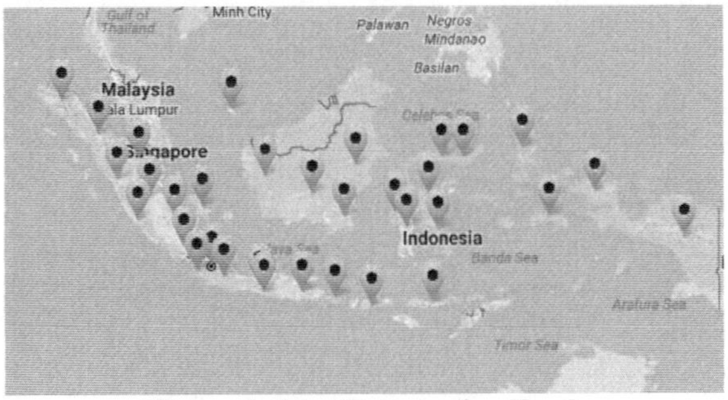

Figure 1. Map Output using K = 2

While figure 2 shows map output using K = 3.

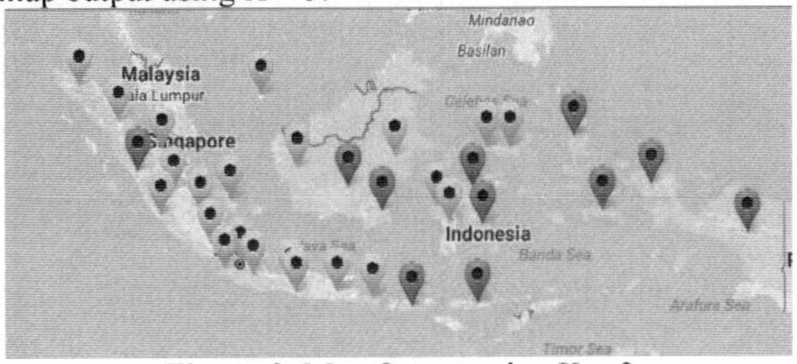

Figure 2. Map Output using K = 3

5. Conclusion

From the above problems it can be concluded that the application of web-based data mining with k-means algorithm can be used to cluster provinces in Indonesia based on primary cooking fuel in household. With this web-based application, it will facilitate government to determine the characteristic of a province in the use of cooking fuel. The application has map output to give easy first glance on the cluster along with the table to give more detail characters. This work can be enhanced using time-series data stream, as each province might have different percentage from each year survey.

Acknowledgment

This work was supported by Universitas Islam Indonesia's internal research funding.

6. References

[1] Kementrian ESDM (Energi dan Sumber Daya Mineral) 2011 Handbook of Energy & Economic Statistics of Indonesia. http://prokum.esdm.go.id/ Publikasi/Handbook%20of%20Energy%20&%20Economic%20Statistics %20of%20Indonesia%20/Handbook %202011.pdf.

[2] Rahmani P, Hartono D M and Kusnoputranto H 2013 Kajian Kelayakan Pemanfaatan Biogas dari Pengolahan Air Limbah untuk Memasak *J. Ilmu Lingkungan* Volume 11 Issue 2: 132-40.

[3] Ashok K D and Kannathasan N 2013 A Study and Characterization of Chemical Properties of

Soil Surface Data Using K-Means Algorithm *Proc. of Pattern Recognition, Informatics and Medical Eng. (PRIME)*

[4] Hot E and Popović-Bugarin V 2015 Soil Data Clustering by Using K-Means and Fuzzy K-Means Algorithm *Proc. of 23rd Telecommunications forum TELFOR 2015*

[5] Kauffman L and Rousseeuw P J 1990 *Finding Group in Data: An Introduction to Cluster Analysis* (New York: Wiley).

[6] Ashok Kumar D, Annie M C L C, and Begum T U S 2012 Computational Time Factor Analysis of K-Means Algorithm on Actual and Transformed Data Clustering *In Proc. of Pattern Recognition, Informatics and Medical Eng. (PRIME),*

[7] Macqueen J 1967 Some Methods for Classification and Analysis of Multivariate Observations *Proc. of the 5th Berkeley Symposium on Mathematical Statistics and Probability* Volume 1 (Berkley C.A.: University of California Press) pp 281-97.

[8] Jain A and Dubes R 1988 *Algorithms for Clustering Data Englewood Cliffs* (NJ: Prentice Hall).

[9] Gan G, Wu J and Ma C 2007 Data Clustering: Theory, Algorithms, and Applications *ASA-SIAM Series on Statistics and Applied Probability* SIAM Philadelphia ASA Alexandria VA

[10] Rousseeuw P J 1987 Silhouettes: a Graphical Aid to the Interpretation and Validation of Cluster Analysis *J. of Computational and Applied Math.* **20** pp 53-65

Hierarchical Clustering for Development Equality Analysis: Indonesian Data of Educational Support (2011 – 2014)

Feri Wijayanto

Department of Informatics, Faculty of Industrial Technology, Universitas Islam Indonesia

feri.wijayanto@uii.ac.id

Abstract. Indonesia which contains more than 30 provinces with the decentralization system needs to identify its development equality. Because inequality in development performance will bring disparity among provinces. At present, the development monitoring is using the indicator's values and comparing those values among provinces. There are no tools which could be used to identify the general development performance, moreover regarding the equality. This research wants to see the possibility of using hierarchical clustering to observe this equality, especially on educational support development. In result, the graph which is plotted using the dissimilarity values as a side result of hierarchical clustering could describe the trend of the equality.

1. Introduction

Indonesia is a vast country which contains more than 30 provinces which are separated by the ocean because Indonesia is an archipelago. It will need a ton efforts from the central government when all of the development is centralized. But on contrary, because of the implementation of decentralization system, the development performance need to be monitored and evaluated since every province has different problem and their development planning will not be the same to each other.

At present, the development performance is indicated by a pack of indicators. With the condition, the monitoring process is conducted by comparing the indicator's value of every province. Those indicator's values are published annually by the Central Bureau of Statistics or related ministry. Even though the values could be compared, the general performance of each province could not be compared [1]. Moreover, the general equality of the development of all provinces could not be identified. Start from that problem this research will try to use hierarchical clustering to evaluate the development equality in Indonesia.

The usage of clustering method is coming up along with idea that the terminology of "equality" is assumed to be the same with "similar". It means that the performances are "equal" among all of objects if they are "similar" to each other. The clustering method which is developed by using "dissimilarity" concept which is assumed to be the same with the term of "inequality". So by using the dissimilarity concept along with clustering method, it could be inferred that the lower the dissimilarity, it would higher in equality. Based on those backgrounds, this research plans to observe whether the dissimilarity concept in clustering method could be used to monitor the development equality, especially in educational support development.

2. Methodology

Basically, the methodology of this research, the analysis of development equality, is based on the clustering analysis. The clustering process need to be conducted before the dissimilarities could be obtained. The dissimilarity values are projected to be fetched from the clustering process because by implementing the clustering the similar objects could be grouped into a cluster and the other similar objects into another cluster. In the end, the similarity of an object relative to the other object could be concluded. Based on the research objective, the hierarchical clustering is selected to be used in this research. The usage of the hierarchical clustering due to the possibility of using dendrogram [2] to represent the creation process of cluster which is equipped with the dissimilarity values of each sequence.

In general, the methodology which are used in this research are shown in **Figure 1**. The first stage of this research is collecting data which are related to Indonesia educational support in four years start from 2011 to 2014. After the collection process, the clustering methodology is implemented, starting with data preprocessing which continues with normalization and transformation [3]. The stage then moved on computing the optimal number of cluster and the clustering process itself. In this research, the important results which are expected from the clustering process is not the cluster, but the distances which occurs on the cluster creation process. Then the distances are collected and represented as a graphic. Analysis and interpretation to the graphic need to be conducted in the last step in order to conclude the hierarchical clustering possibility to observe the development equality.

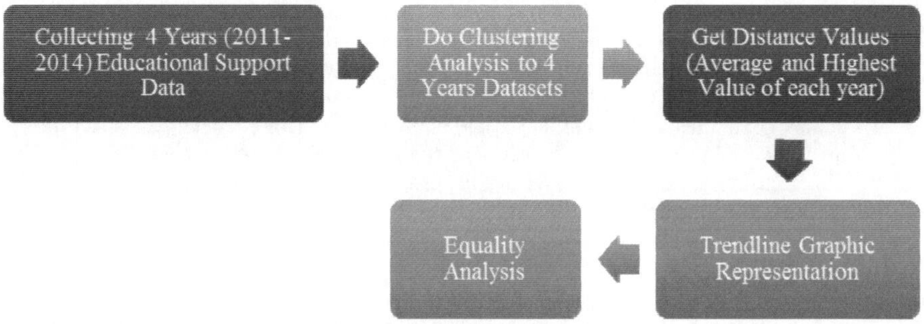

Figure 1. Research Methodology

2.1. *Educational support dataset collections*

The data is collected from the Senior Secondary School Statistic which is published by The Educational Data and Statistic Center, Ministry of Education and Culture. While the dataset is formed based on the input – support – output – context model which have been proposed in previous research [4] which is shown on **Figure 2**. In previous research on educational effectiveness, it is proposed to use an input – process – output – context to measure the educational development effectiveness. Basically this model is adapted from Scheerens' model [5] which using the OECD indicators. So, based on the model the educational support dataset will be formed by using the six indicators in the context component.

2.2. *Clustering analysis*

The clustering algorithm which will be used in this research is the simplest because the focus of the discussion here is not the resulted cluster but rather on the distance values which are produced in conjunction with clustering process. So, the hierarchical clustering method is used in this research. Another reason of the method usage is due to the requirement of distances tracking for each sequence of cluster creation. Moreover, the hierarchical clustering, especially complete link, has the ability to focus on outliers [6] which means it becomes possible to identify the unusual phenomenon.

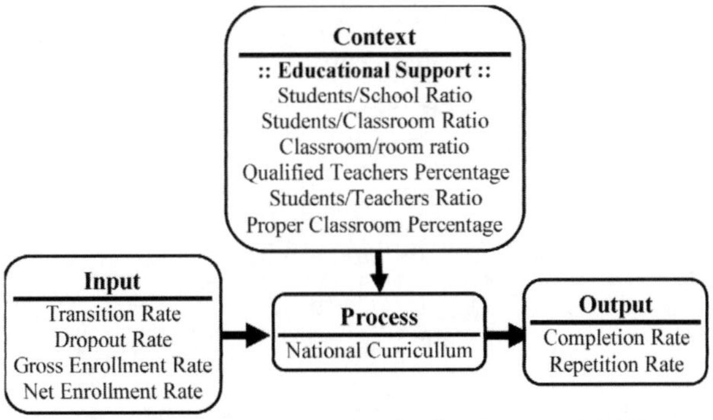

Figure 2. The proposed effectiveness model [4]

On general, the common hierarchical clustering methodology has been used. The educational support dataset which has six features like on Table 1 is normalized and transformed before it clustered. The min-max normalization[2] and cost-benefit analysis[7] transformation is implemented on the preprocessing. The methodology for clustering analysis could be described in **Figure 3**. Basically, on clustering analysis, number of clusters is an important information but here, the step is excluded because the goal here is only obtaining the distance of the agglomerative process.

Table 1. Educational Support features' dataset

No.	Feature
1	Students/School Ratio
2	Students/Classroom Ratio
3	Classroom/room ratio
4	Qualified Teachers Percentage
5	Students/Teachers Ratio
6	Proper Classroom Percentage

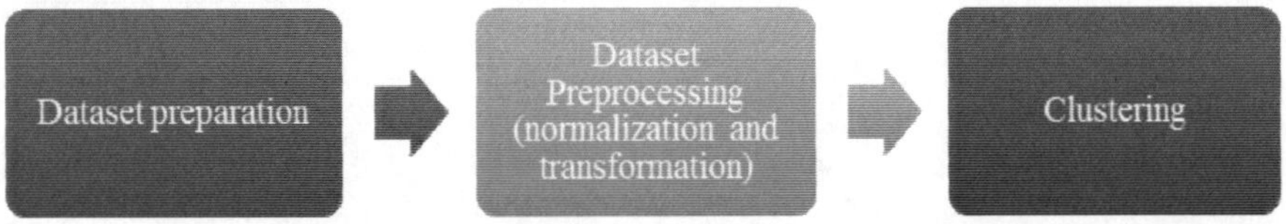

Figure 3. Clustering analysis methodology

2.3. Distance values collections

One other thing that makes hierarchical clustering is implemented on this research is dendrogram [8]. The cluster creation on hierarchical clustering is having the capability to be represented as a dendrogram. Dendrogram has parameter which is needed for this analysis that is known as height, as represented on **Figure 4**, **Figure 5**, **Figure 6** and **Figure 7**. Height is parameter which represents the distance values which are computed for every agglomeration of an object. The lower the distance the more similar it should be and the more equal it becomes. Those heights are then to be collected in order to be graphed on a trendline graph in agglomeration process sequence order which is shown in

Figure 10. Further than that, the average of those heights and the highest heights in four years are taken in order to be graphed on **Figure 8** and **Figure 9**.

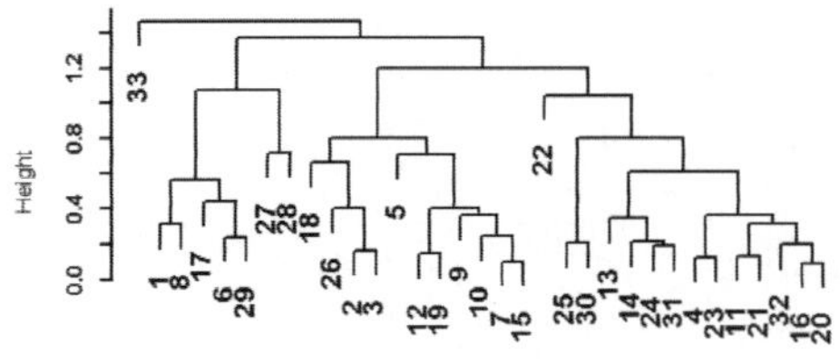

Figure 4. Dendogram of 2011 educational support dataset

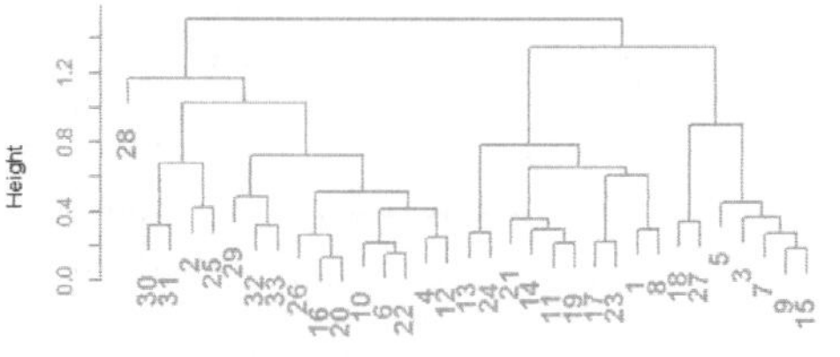

Figure 5. Dendogram of 2012 educational support dataset

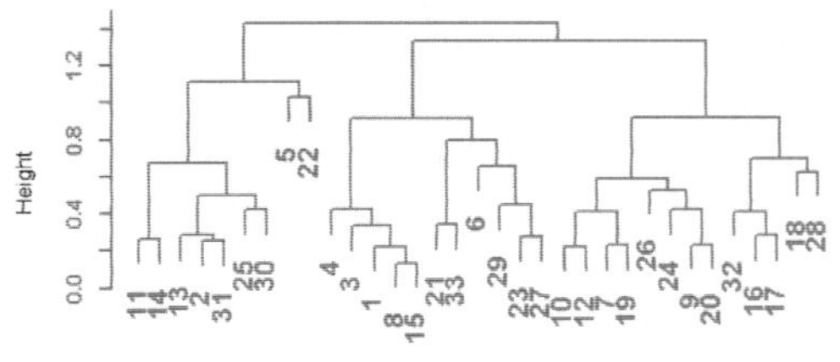

Figure 6. Dendogram of 2013 educational support dataset

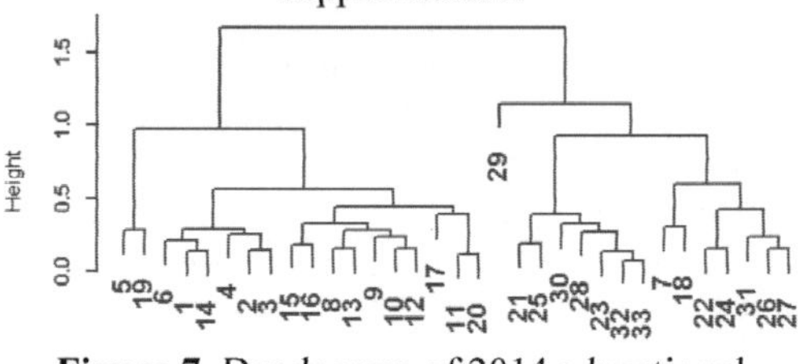

Figure 7. Dendogram of 2014 educational support dataset

3. Results and Discussions

After the distances, based on the dissimilarities from the clustering process, are collected, they are then plotted on graphs which are observable on the **Figure 8** and **Figure 9**. **Figure 8** shows the trend of dissimilarities by plotting the highest value of computed distance (last computed distance) from each year dataset while **Figure 9** forms the trend by plotting the computed average of distance values from each year dataset. By comparing **Figure 8** and **Figure 9** it is easy to conclude that the trend is opposite. **Figure 8**'s trend is relatively increasing while **Figure 9** is relatively declining. Basically, it could be simply inferred that on average the dissimilarities is declining but on contrary there are provinces which have quite different performance on educational support provision.

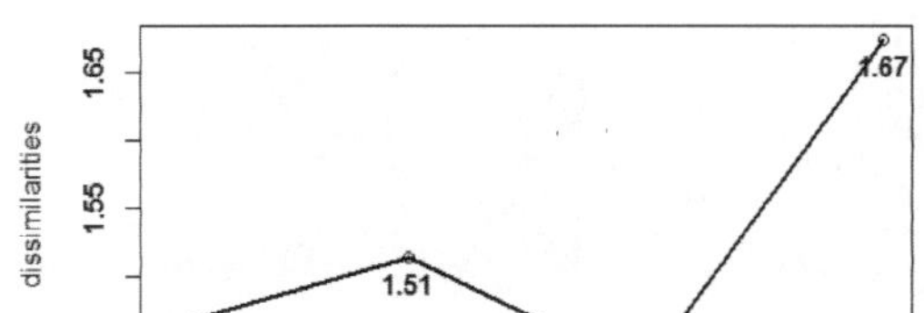

Figure 8. Trend of highest dissimilarity values

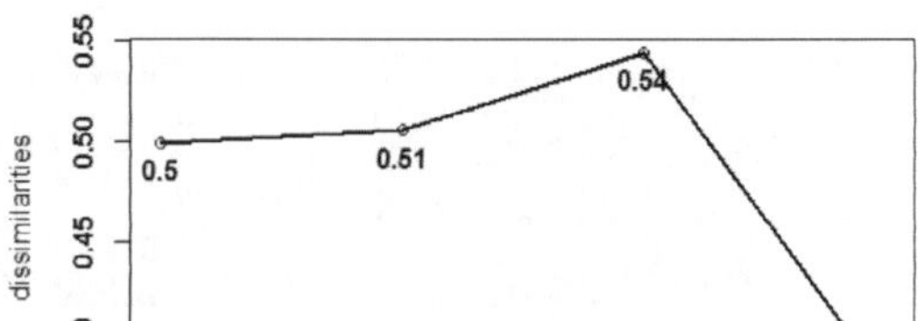

Figure 9. Trend of average dissimilarity values

Theoretically, equality in educational development could be obtained if the dissimilarities are zero or approaching zero, but that is impossible. Actually, if it could be determined how near to zero of dissimilarities which could be said that is still considered equal, the trendline graph on **Figure 8** and **Figure 9** would be rich in information regarding the equality.

The most used distance formula in clustering computation is the Euclidean as written in equation (1). Based on the equation (1), if the dataset contains six parameters the highest value of distance which could be obtained is six. It is assumed that the datasets are normalized into 0 to 1. Thus, in general, what could be said on the dissimilarities by referring to **Figure 8** and **Figure 9** is that the average dissimilarities is quite low that is only about 0.5 or 8% to the highest value possible for the first three years and 0.38 or 6,7% for 2014.

$$D_{ij} = \left(\sum_{l=1}^{d} (x_{il} - x_{jl})^2 \right)^{1/2} \quad (1)$$

where:

D_{ij} = The distance between object i and j
x_{il} = The lth parameter of object i
x_{jl} = The lth parameter of object j

The results on the **Figure 8** and **Figure 9** could be clarified with **Figure 10** which shows the trend of dissimilarities on the sequence of cluster creation in four years (2011-2014). For the highest value (last value) of dissimilarity in each year is 2014, 2012, 2011 and 2013 respectively from the highest value to the lowest. It matches with trend on **Figure 8**. If the **Figure 10** is closely observed, it could be predicted although 2014 has the highest dissimilarity on the last cluster creation on contrary averagely 2014 has the lowest dissimilarity.

So basically, by analyzing the **Figure 8** and **Figure 9** it could be inferred that the graph could be used to describe the equality trend in a certain range of time. In this research case, it could be inferred that in general, the educational support development performance among provinces in Indonesia is becoming more equal year by year from 2011 to 2014 but on contrary the highest dissimilarity in general become higher.

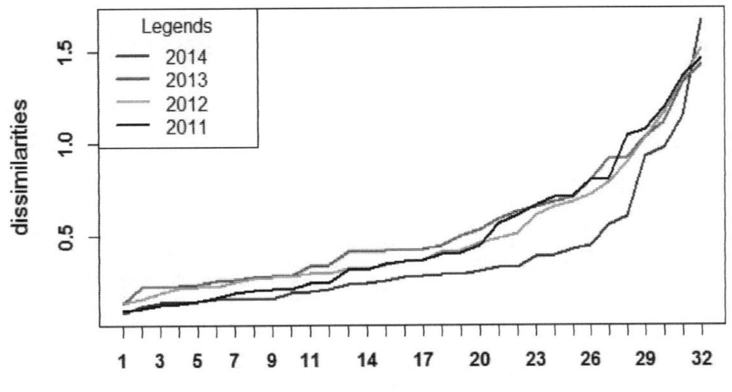

Figure 10. Dissimilarities trend of cluster sequence in four years (2011-2014)

4. Conclusion

In conclusion, the plotted dissimilarities could be used to observe the growth of the dissimilarities year by year. The growth of dissimilarities, if compared to the definition, would means the decline of equality. On contrary, the equality would be obtained if only the dissimilarity is zero. Even though it would be impossible, it could be simply said that the lower the dissimilarity would be higher in equality.

Using the graph behavior, the equality of educational support development could be observed and deduced. Observing on average dissimilarity trend, it could be concluded that in general the performance of educational support development becoming more equal. While as a supplementary, by using the highest dissimilarity value trend, it could be alerted that there are any extraordinary cases or outliers in some provinces.

5. References

[1] Wijayanto F Input-Support-Output Model Evaluation Using Clustering Analysis on Indonesia High School Dataset *Adv. Sci. Lett. (forthcoming)*

[2] Han J, Kamber M and Pei J 2006 *Data Mining: Concepts and Techniques* (San Francisco: Morgan Kaufmann Publishers Inc.)

[3] Berkhin P 2006 A Survey of Clustering Data Mining Techniques *Grouping Multidimensional Data* (Berlin/Heidelberg: Springer-Verlag) pp 25–71

[4] Wijayanto F Clustering Analysis on Indonesian Education Quality Performance Using Input-Output Model *Adv. Sci. Lett. (forthcoming)*

[5] Scheerens J, Luyten H and van Ravens J 2011 Measuring Educational Quality by Means of Indicators *Perspectives on Educational Quality* (Springer Netherlands) pp 35–50

[6] Madhulatha T S 2012 An Overview on Clustering Methods *IOSR J. Eng.* **2** 719–25

[7] Larose D T 2005 Discovering Knowledge in Data: An Introduction to Data Mining (John Wiley & Sons)

[8] Kotsiantis S B and Pintelas P E 2004 Recent Advances in Clustering : A Brief Survey *Methods* **1** 73–81

Network gateway security method for enterprise Grid: a literature review

A Sujarwo[1], J Tan[2]

[1] Department of Informatics, Universitas Islam Indonesia, Yogyakarta, Indonesia

[2] Faculty of Information Technology, Monash University, Caulfield, Australia

E-mail: ari.sujarwo@uii.ac.id, jefferson.tan@monash.edu

Abstract. The computational Grid has brought big computational resources closer to scientists.It enables people to do a large computational job anytime and anywhere without any physical border anymore. However, the massive and spread of computer participants either as user or computational provider arise problems in security. The challenge is on how the security system, especially the one which filters data in the gateway could works in flexibility depends on the registered Grid participants. This paper surveys what people have done to approach this challenge, in order to find the better and new method for enterprise Grid. The findings of this paper is the dynamically controlled enterprise firewall to secure the Grid resources from unwanted connections with a new firewall controlling method and components.

1. Introduction

A *computational Grid* is a facility for people to do massive computing over a uniformly shaped infrastructure of diverse resources [1]. Many resources from network-connected physical or virtual computers work together within the Grid infrastructure to provide a massive system. This system is designed to process multiple jobs that need intensive computation. Certain protocols are used to organize many computational nodes and facilitate the hosting of jobs that are submitted into them.

As a massive system that consists of various parts such as computational components, shared storage and many other components, Grid infrastructures are requiring advanced security. Some of the Grid resource components and data are potential for attack. The sensitive data and the computation resources must be secured properly to avoid unauthorized access. As an example, Foster and Kesselmen [2] explained in his book that computational and experimental scientists and engineers use the Grid to support their works. Data are saved and processed within the Grid resources. When these data are not secured properly, it could attract unauthorized people to access and possibly spread restricted information. In certain cases, the consequences of this issue could be damaging. Some requirements are already specified to provide basic and advanced security to this infrastructure to achieve better security for it.

2. Grid Protocols

Grid resources is controlled by the function of each Grid protocol layer. [3] defined the protocol stack of a Grid architecture that is shown in the left hand side of Figure 1, Grid consists of five layers: fabric, connectivity, resource, collective, and application.

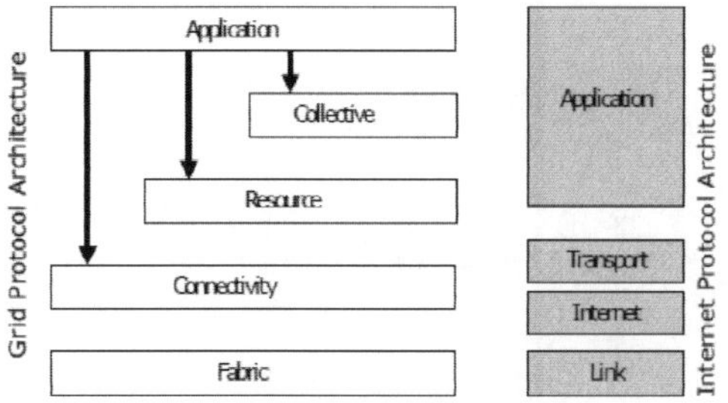

Figure 1 Grid architecture related to Internet Protocol architecture, borrowed from [3]

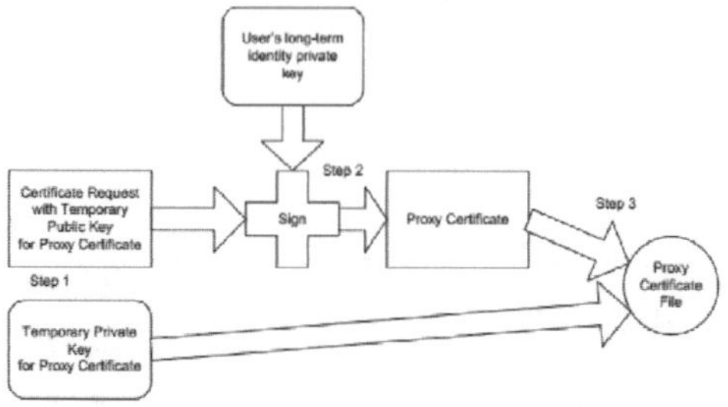

Figure 2 Proxy certificate for single sign-on, borrowed from [4]

The functionality of each layer related to the Internet Protocol (IP) architecture in the right hand side of Figure 1. The Fabric layer consists of resources owned by each node such as storage and computational resources that are shared within Grids. All of them are then interconnected using a Grid protocol. In regard to the IP architecture, the fabric layer related to the IP link layer in terms of providing the necessary hardware to be used by the upper layer. Among all of those layers, the Grid middleware such as Globus located in two layers, which is the Connectivity layer and the Resource layer [5]. These layers are closely related on how the Grid middleware provides the security and resource management of the Grid.

3. The Challenge

The Grid has characteristics that is called "the nature of the Grid" [3] which is the characteristics of users, resources, policies, and other components of a Grid. Each of those components tends to be dynamic. The number of users and organizations joining Gris as Resource Provider can be changing all the time. User at any time can join and submit a certain task. The Grid infrastructure scopes are typically wide. It could be within a university, group of the education system, or it could be extensive for certain purposes. The users could come from worldwide. As long they have an access to the node joining the Grid, they would able to run any tasks, anytime, anywhere. Furthermore, the number of resources can vary. A real organization is possible to add or withdraw its resources from a Grid. Sometimes the participating resources facing a problem and need to be turned off, it makes the number of resources possible to change.

While requesting a resource, users are likely to be assigned to an available resource and need a certain amount of time to submit tasks, distribute the data among nodes and the execution of the tasks. It creates a variable time between each task in regards to starting time and ending time. So the requirements of the Grid resources are likely to vary and make the computation cycle more dynamic. The nature of Grid mentioned above brings a consequence in term of security. The security system must be able to overcome the Grid infrastructure requirements to keep Grid running well without compromising the whole system security.

4. Basic Grid Security

The Grid Security Infrastructure (GSI) is an integrated security system for Globus which provides the basic security infrastructures for Grids [6]. The GSI mainly focus on the phase of user verification [7]. The user verification is the primary security function that Grids must have to identify users and offer services to users. Based on [8], the basic security of a Grid requires Security Policy as necessary rules to define the involved objects to the Grid, The inter-organization security that covers the security among different organizations either internal or external. There is no doubt that disruption can come from anywhere even from internal organization. Grids also need a larger security system to protect from external disruption. Authentication and authorization mechanism are designed with properties that all of the Grid components remain safe and running properly. These are two Grid's basic security: Grid User Credentials – The X.509v3 Certificate and Proxy Certificate for Single Sign-on (Figure 2).

5. Approaches

The explanation about the Grid architecture and the basic security of a Grid system gave a clear understanding that the Grid needs more advanced security system. Many security schemes are invented for a common network and server, such as the authentication system and the firewall. The useful way to protect a Grid is separating the Grid infrastructure from the public network and placing the Grid system behind a firewall. Newman [9] have defined the firewall as an entity that has a right to enforce rules to make a decision on passing data over a network. The firewalls are applied to filter all incoming and outgoing data to the Grid. However, the nature of firewalls is static. Static firewalls need administrators to add or remove the firewall rules. This nature of the firewalls is contradicted with the nature of Grid which is dynamic.

The Grid needs several ports to be opened in order to operate. The needed ports are depending on each user's requirements. To allow a user to access the Grid, certain firewall rules are added to the firewall. Since the firewall is not controlled all the time, it makes rules stay active on the firewall longer until the administrators change or remove the rules. The unused ports remain accessible for a long time; this situation makes the Grid infrastructure more vulnerable. To overcome this problem, scientists and researchers have tried to find out the better way to operate a Grid in a firewalled area with the integration of the currently invented security scheme. There are two approach to overcome this problem: Proxy mechanism and dynamic firewall.

5.1. Proxy Mechanism for Firewall Traversal

The concept of proxy mechanism is by operating a currently opened firewall ports to establish a new connection between the users and resources [10]. An application sits between the users and the Grid infrastructure to accept the user's connection, receive data, and forward it to the Grid. The following explains several methods on how the proxy mechanism works to move data through the firewall. Metsch [11] explained *Generic Connection Brokering (GCB)* as a the first technique to enable data passing firewall by relaying the data packets between the users and the Grid. As shown in Figure 3, the left hand side is the initiating party (could be an external network). It sends data to a relay point which is a *GCBbroker*. The broker then makes a decision to allow or reject the connection based on the firewall rules.When the connection is allowed, the GCB then negotiates a connection to the Grid infrastructure, and let the Grid to send a reply to the client through a relay point.

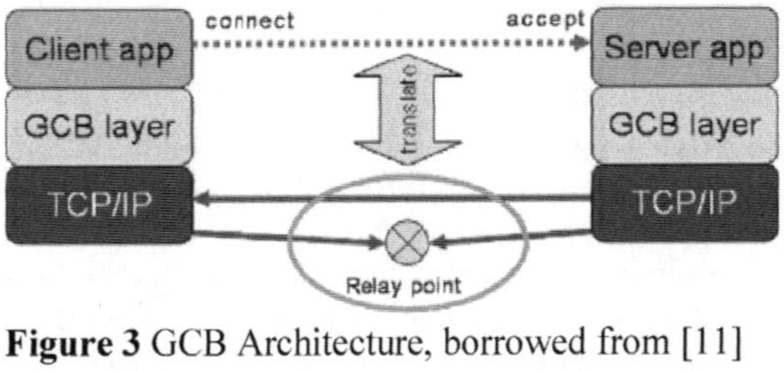

Figure 3 GCB Architecture, borrowed from [11]

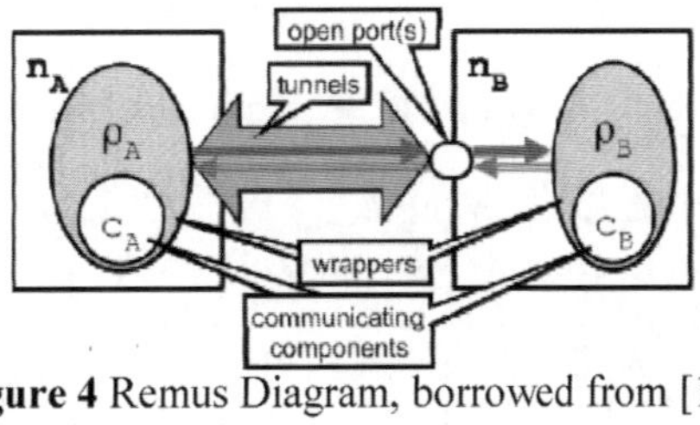

Figure 4 Remus Diagram, borrowed from [12]

Considering to this project design which avoids an excessive change to the Grid infrastructure or the client's application, the GCB is not an appropriate approach to be applied in this project. The GCB needs a modification either in the Grid infrastructure and the client to insert a GCB layer. This layer is a library that is invoked when the application wants to send a message to the GCB broker[11].

The second technique called *Remus*. Rerouting and multiplexing system are the way *Remus* works. [13] created Remus to pass data across a firewall by operating the currently registered port within a firewall. All data are wrapped as new packets addressed with port numbers that are open in the firewall. The Remus diagram is shown in Figure 4. Remus approach requires data to be processed in additional processes. The data are encapsulated several times to be able to pass the tunnel. [12] explained that this process increases the size of the data, and makes data heavy enough to be transferred. The Remus is not suitable for this project implementation model caused by the heavy data size, but it gives an understanding that the proxy and tunnel mode put a large data overhead to be able to traverse a firewall.

5.2. Dynamic Firewall

Another approach to traverse a firewall besides proxy mode is by managing a firewall and makes a firewall more dynamic to modify the rules without any administrator intervention. This method enables a firewall to alter its rules when user is successfully authenticated such as *Cooperative On-DemandOpening (CODO)* –a method to change firewall rules based on user request, *Dyna-fire* –works by theknocking mechanism, *Firewall Traversal Protocol (FiTP)*, and *Romulus*. Cooperative On-Demand Opening (CODO) works by planting a *CODO agent* inside the firewall [11]. This is called the *CODOfirewall agent (FA)*. CODO also requires a *CODO client library (CL)* to be integrated in each of user'sapplication that wants to pass through a firewall [14] as shown in Figure 5.

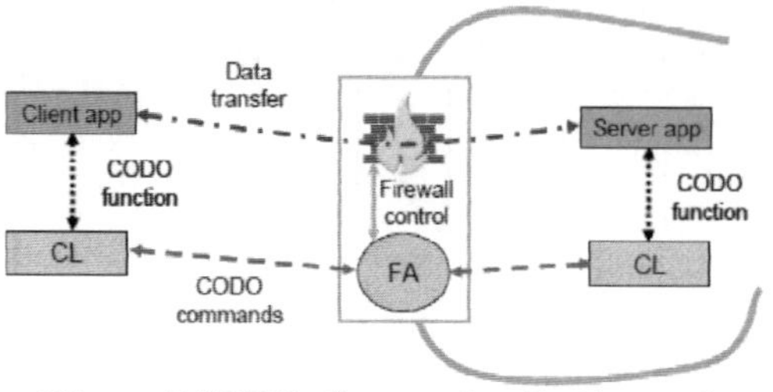

Figure 5 CODO: Cooperative On Demand Opening, borrowed from [14]

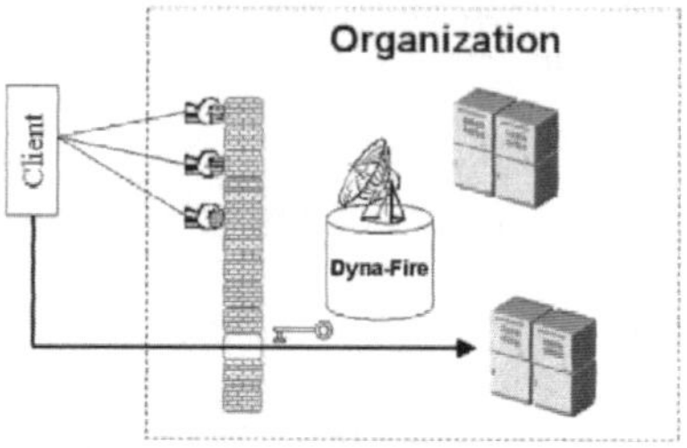

Figure 6 Dyna-fire port knocking, borrowed from [15]

The CODO needs its agent installed within the communicating devices. For certain systems, it would not be feasible to add an application to it. Rewriting all the client's application and the Grid infrastructure is not a simple matter, so another approach with a simpler solution is required. Almost the same with *CODO,* Dyna-Fire use knocking to open a firewall port. It needs the *Dyna-fire daemon* to be installed within a firewall. [15] explained that four aspects must be fulfilled to be allowed to pass data: client must encrypt the knock in a right encryption method; client must ensure that the format of knocks sequence is right; client must use registered IP address; and a valid user's identity must be presented. A set of firewall rules are be added to open firewall ports after the client fulfils those requirements as shown in Figure 6. However, a problem with Dyna-fire is that the Dyna-fire is possible to be attacked with a *replay* and a *man-in-the-middle attack* [11]. The next approach is called Firewall Traversal Protocol (FiTP) released by Firewall Virtualization for Grid Applications – Working Group (FVGA-WG), a group that works focus on proposing a security solution for a Grid based on the current firewall filtering system. The FiTP plan was introduced in 2009 [16]. Up to today, there is no developed protocol that manages a firewall directly to make a firewall aware of what kind of application data needs to traverse the firewall. The FiTP is proposed to be a protocol that has such ability. The FiTP diagram is shown in Figure 7.

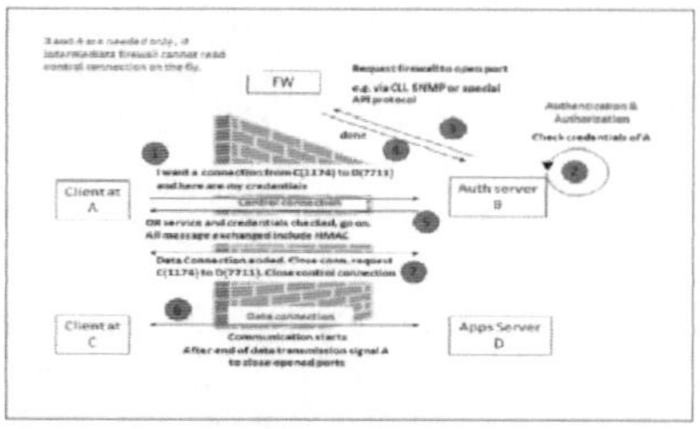

Figure 7 FiTP diagram, borrowed from [17]

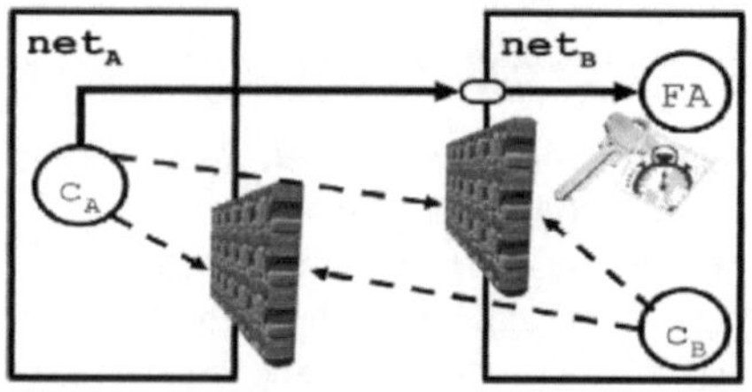

Figure 8 Romulus pinhole request process, borrowed from [13]

Since FiTP has not been done, [18] introducing Romulus as a controller to open and close firewall with the use of agent. The firewall agent has a right to accept a request sent by users or reject the request if it does not satisfy the security requirements. It runs a security procedure to ensure the incoming message was sent from a valid user. It authenticates user based on the validity of the use certificate [13] as shown in Figure 8. [13] implemented the Romulus in two type of firewalls: a Linux IPTables and a Cisco router Access Control Lists (ACL). For a larger scope, based on research done by [19], IPTables do not have the capabilities to check all the incoming packet coming from a large network due to a hardware and software constraints of the PC firewall. The implementation of the IPTables and the Cisco ACL as the additional security system within a Grid infrastructure could make Grid running underneath the performance standard due to the capability constraints in both of those firewalls. To avoid the potential problem, the Grid infrastructure supposed to use a professional stand-alone firewall [20] as the enterprise security.

6. Evaluation and Conclusion

The Grid infrastructure interconnects computers to cooperate in such a way to finish tasks submitted to the Grid. As a large system comprises of components, the Grid needs an advanced additional security system. The firewall proposed as an additional security system to filter all the incoming data. The firewall becomes the potential security system for a Grid. However, some issues such as the static rules of the firewall could arise security vulnerability for Grid. The Romulus could be a model of the firewall controlling method for enterprise Grid infrastructure. The method used by Romulus that is called firewall virtualization, could be extended with the use of authentication, authorization, and accounting to give a more flexible credential checking to ensure the accessing party is a valid one.

References

[1] Foster I, Zhao Y, Raicu I and Lu S 2008 Cloud Computing and Grid Computing 360-Degree Compared *2008 Grid Comput. Environ. Work.* pp. 1–10.

[2] Foster I and Kesselman C 2003 *The Grid 2: Blueprint for a new computing infrastructure* (Elsevier).

[3] Foster I, Kesselman C and Tuecke S 2001 The anatomy of the grid: Enabling scalable virtual organizations *Int. J. High Perform. Comput. Appl.* **vol. 15** no. 3 pp. 200–22.

[4] Welch V, Foster I, Kesselman C, Mulmo O, Pearlman L, Tuecke S, Gawor J, Meder S and Siebenlist F 2004 X. 509 proxy certificates for dynamic delegation in *3rd annual PKI R&Dworkshop* **vol. 14**.

[5] Foster I 2003 The grid: A new infrastructure for 21st century science *Grid Comput. Mak. Glob.Infrastruct. a Real.* pp. 51–63.

[6] Butler R, Welch V, Engert D, Foster I, Tuecke S, Volmer J and Kesselman C 2000 A national-scale authentication infrastructure *Computer (Long. Beach. Calif).* **vol. 33** no. 12 pp. 60–6.

[7] Foster I and Kesselman C 1999 The Globus project: A status report *Futur. Gener. Comput. Syst.* **vol. 15** no. 5 pp. 607–21.

[8] Foster I, Kesselman C, Tsudik G, and Tuecke S 1998 A security architecture for computational grids in *Proceedings of the 5th ACM conference on Computer and communications security* pp. 83–92.

[9] Newman D 1999 *Benchmarking terminology for firewall performance*.

[10] Mahjoubi P 2010 *RADIUS-based Firewall Virtualization for Grids* (Monash University).

[11] Metsch G I T, Gommans F R L, de Smet A, Volpato G L, and LUH R 2010 *Requirements on operating Grids in Firewalled Environments*.

[12] Tan J, Abramson D and Enticott C 2009 REMUS: a rerouting and multiplexing system for Grid connectivity across firewalls *J. Grid Comput.* **vol. 7** no. 1 pp. 25–50.

[13] Tan J, Abramson D and Enticott C 2009 A virtual connectivity layer for grids in *Fifth IEEE International Conference on e-Science* pp. 307–12.

[14] Son S, Allcock B and Livny 2005 M Codo: Firewall traversal by cooperative on-demand opening in *Proceedings. 14th IEEE International Symposium on High Performance Distributed Computing HPDC-14* pp. 233–42.
[15] Green M L, Gallo S M and Miller R 2004 Grid-enabled virtual organization based dynamic firewall in *Proceedings of the 5th IEEE/ACM International Workshop on Grid Computing* pp. 208–216.
[16] Niederberger R 2009 FiTP-A protocol draft for dynamic opening of Firewalls in *25th Open Grid Forum (OGF25)*.
[17] Niederberger R, Monga I and Metsch T 2008 Firewall Virtualization for Grid Applications in *Open Grid Forum - OGF23* (Barcelona).
[18] Tan J, Abramson D, and Enticott C 2010 Firewall Traversal in the Grid Architecture in *12th IEEE International Conference on High Performance Computing and Communications (HPCC)* pp. 189–96.
[19] Ranganath V P and Andresen D 2003 A set-based approach to packet classification in *Proceedings of the IASTED International Conference on Parallel and Distributed Computing and Systems* pp. 889–94.
[20] Thomason S 2012 Improving Network Security: Next Generation Firewalls and Advanced Packet Inspection Devices *GJCST-E Network, Web Secur.* vol. **12** no. 13.

The Development of Smart Home System for Controlling and Monitoring Energy Consumption using WebSocket Protocol

Niti Witthayawiroj[1] and Pongpon Nilaphruek[2]
[1,2] Department of Mathematics and Computer science, Faculty of Science
Rajamangala University of Technology Thanyaburi 39 Moo1, Rangsit-Nakhonnayok Rd. Klong6, Thanyaburi Pathum Thani, Thailand

E-mail: [1]niti@rmutt.ac.th, [2]pongpon_n@rmutt.ac.th

Abstract. Energy consumption especially electricity is considered one of the most serious problems in households these days. It is because the amount of electricity consumed is more than the amount that people actually need. This means that there is an overusing which resulted from the inconvenience of moving to the switch to turn off the light or any appliances and it is often that closing the light is forgettable, for instance; in addition, there are no tools for monitoring how much energy that is consumed in residents. From this, it can be easily seen that people are having a problem in energy usage monitor and control. There are two main objectives of this study including 1) creating the communication framework among server, clients and devices, and 2) developing the prototype system that try to solve the mentioned problems which gives the user an opportunity to know the amount of electricity they have used in their houses and also the ability to turn appliances on and off through the Internet on smart devices such as smart phones and tablets that support Android platform or any web browser. Raspberry Pi is used as a microcontroller and the data is transferred to the smart device by WebSocket protocol which is strongly recommended for real-time communication. The example features on the device's screen are user management, controlling and monitoring of appliances. The result expresses that the system is very effective and not difficult to use from users' satisfaction. However, current sensors may be used for a more accurate electricity measurement and Wi-Fi module for more appliances to calculate its power in the future.

Keywords- Smart Home; WebSocket protocol; Monitoring; Controlling; Internet of Things

1. Introduction

Internet has been an important technology for today and undoubtedly the future. This is because it erases the border of cyber world between countries and each parts of the world. The statistic from Internet World Stats [1] has shown that the use of the Internet all around the world has been dramatically increasing shown in Figure 1. The ability to connect wirelessly has brought about the term "Smart Home", the system that provides convenience and safety in residents. A smart home system or home automation is usually based on Wi-Fi in terms of data transmission between light bulbs and appliances to user's smart devices such as smart phones and tablets; this means that to use a smart home system, smart devices are required for end users with the ability to install applications. The number of world smart phones use after 2013 was more than the use of desktop computer [2]. This trend of Internet use has implied us that people in the world are comfortable with these portable devices and any such device can connect the Internet in the future surely. Smart home comes with that trend, so called "Internet of Things" (IoT), the concept that everything can connect to the Internet giving it the ability to transfer its usage to the user for any purpose. Smart home with IoT together means that any appliances in houses can communicate with the user through the Internet and the information from houses can display on screens of smart devices. Therefore, this study aims to create the smart home system in order to provide the convenience for monitoring the energy consumption in users' home and controlling any electric devices via the Internet.

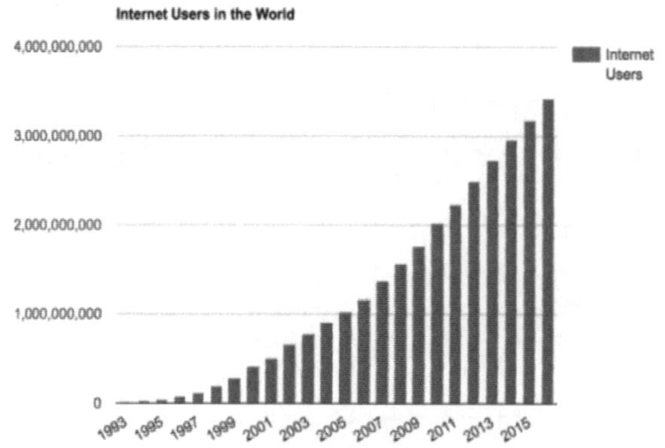

Figure 1. Internet users in the world

2. Related Works

Nowadays, there are many systems supporting the smart home concept. For example, Ramlee and their project mates developed smart home system using Android application [3]; but they use only Bluetooth connection among devices instead over the Internet. Moreover, Kumar developed the smart home system using Android as well [4]. That system uses REST protocol to transfer the data among devices via server. However, the communication among electric appliances and any client devices over the Internet is the big challenge point for researchers. There are many researches try to propose the best communication protocol. Pimentel and Nickerson [5] compared the latency of WebSocket protocol [6] among HTTP polling and long polling [7]. The result of their research shows that the average latency of WebSocket is better than HTTP and long polling. The possible protocols such as CoAP, MQTT, XMPP, RESTFUL, AMQP and WebSocket are strongly recommended for IoT application. In the survey on application layer protocols for the IoT application [8], they indicated that MQTT has proved to be more efficient for battery-run devices but it is not widely used as HTTP, and WebSocket is designed for real-time communication as MQTT but it is mainly used for client/server based architecture. Those protocols are recommended from them. Therefore, this study mainly uses WebSocket protocol for communication framework because the prototype application is based on HTML5, CSS and JavaScript as well as Data Center Unit (called as "DCU") which is the microcontroller connected to the electric appliances supports WebSocket also. In addition, low cost ubiquitous sensing system was tested by [9] and the result shows that the reliability of sensing information transmission through the proposed integrated network architecture is 97%. That means we can apply any low cost sensors to the system with the reliability.

3. Methodology

First, the related works was reviewed in section 2 to acquire the possible solutions for developing the smart home system especially for real-time communication. Second, the system design consists of four parts including:

3.1. Designing Communication Framework

According to the Figure 2, it illustrates the overview of communication framework. The agent of WebSocket was built separately. There are three main parts of WebSocket agents including client, server, and device (call as "DCU").

- First, when some clients wake up, then they will connect with the server using WebSocket agent; after that, do the handshake (see number 1 in the Figure 2).

- Second, when some DCUs wake up, do the same things as clients (see number 2 in the Figure 2). Actually, clients and DCUs must have their own ID to determine that who are connected to the server.
- This time, all active agents can send any message to each other. For example, if a client would like to turn on device 001, it just sends the message such as "please turn on 001" to the server (see number 3 in the Figure 2).
- After that, the server will receive the message. Next, server will send the message to DCU, which connects with device 001. The device 001 will turn on and send responding message back to related clients via the same path to confirm the result (see number 4-5 respectively in the Figure 2).

DCU is used as a microcontroller; in this research, Raspberry PI 2 model B [10] was used to control any electric equipment because it supports WebSocket protocol. Additionally, server WebSocket agent is used as a medium point to control the transmission messages in the system.

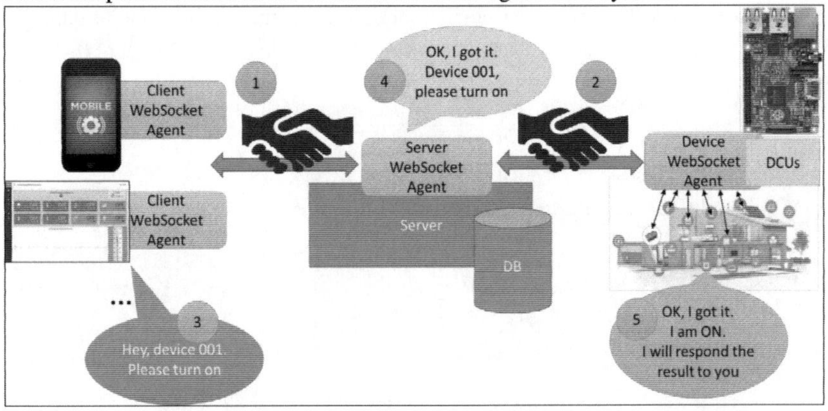

Figure 2. Communication Framework

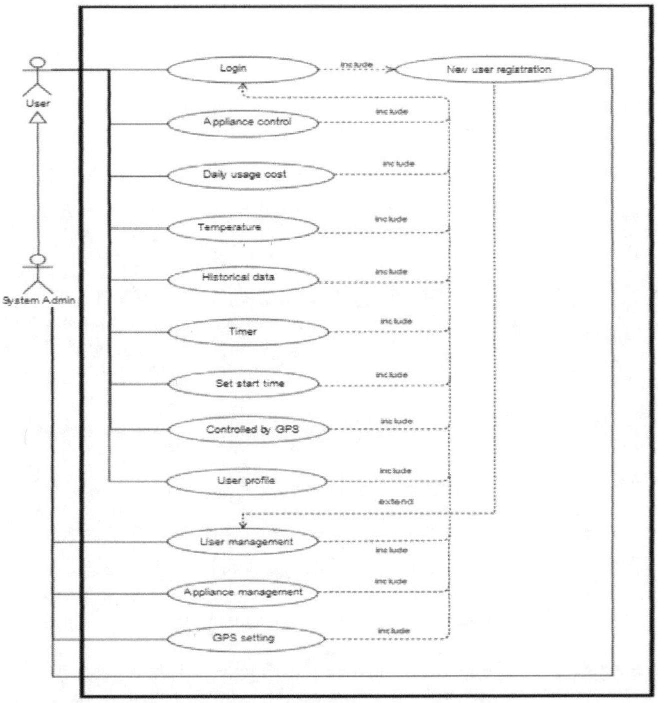

Figure 3. Use Case Diagram of Smart Home System

3.2. Designing System Features

The system features were designed using Use Case Diagram. In the diagram, it shows that which options the user can use and which options are for the system admin. The admin can manage user and appliances while the user cannot, see the Figure 3.

3.3. Designing System Architecture

System architecture has three parts consisting of:

- Client side, user can use any platform to access the application via web browsers. However, this study focuses on Android platform; that means we have the application on Android.
- Server side, all system information is kept in the database on the server as well as we also have rules and metadata to control users who interact with the system.
- DCUs side, it is the microcontroller, which directly connects with electric appliances.

As the mention above, three parts connect together via WebSocket protocol for real-time processing. Furthermore, we also use HTTP request for normal tasks such as manipulating with database, and call Web service that use JSON and XML for data exchange. See the Figure 4 that illustrates the overview of system architecture.

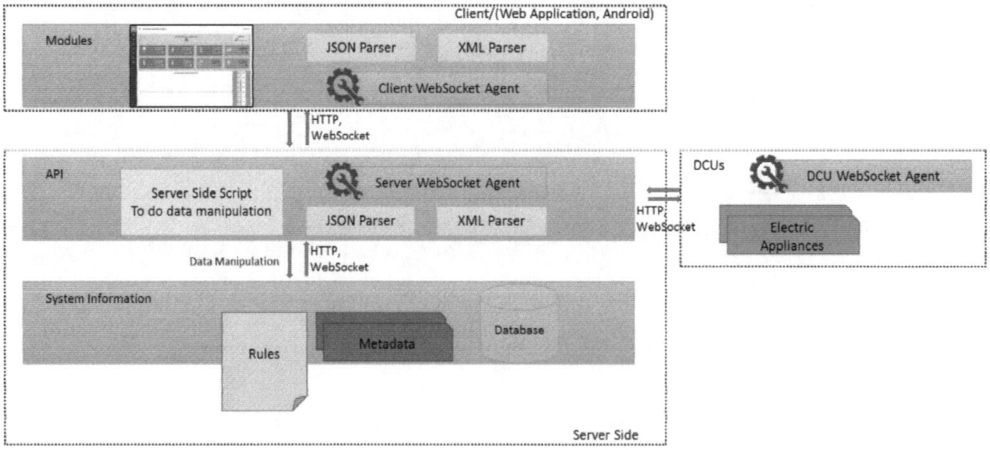

Figure 4. System Architecture

3.4. Constructing Home Model

Figure 5. All appliances on home model, four light bulbs and one fan can be tested on this model

According to the Figure 5, home model was constructed to prove the concept of smart home system in this research. We have four light bulbs, one fan and three electrical outlets. Relay modules are needed in order to act as switches and bridges for working on 220V AC power system in Thai households and 5V DC power system on the Raspberry PI. After the system was completely built, it was tested its performance and we fixed problems that were found such as bug or hardware failure. On the end-user application on android smart phones and tablets, the screen and pages were designed. The result will be explained in the next section.

4. The Prototype System and Result

Once the user has logged in to the system, they are led to the main page called "dashboard". In this screen, all the appliances status is shown, for example, the fan is off and all the three bulbs are as well off. At the top right, the cost of electricity used at that day is calculated real-time and will be shown on this corner as shown in the figure 6 (the right is shown on mobile device).

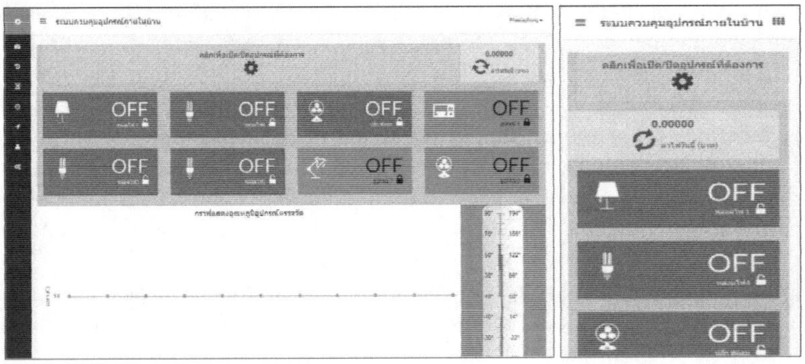

Figure 6. Dashboard page on a normal screen (left) and mobile screen (right)

The figure 7 (left) shows that users can use the GPS setting feature to control electrical appliances by the distance between their location and home location. For example, we can set the distance that if it is far from home more than 3 kilometers, every appliance will be turn off automatically. Moreover, the history of users will be recorded for every transaction as shown in the figure 7 (right) as the mobile screen.

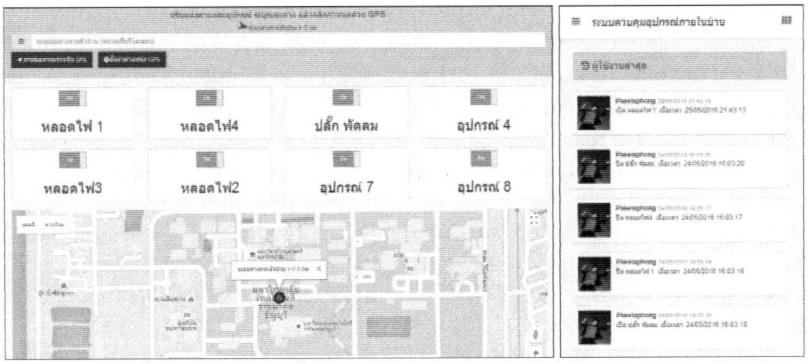

Figure 7. GPS setting (left) and historical usage (right)

In fact, we also have more features such as user management, appliance setting, timer, and calendar setting for controlling electric appliances. The system also was tested on all of features and different devices with different screen resolution by users' satisfaction, and the result express that the average satisfaction from users is in excellent level. Furthermore, the best display of the screen is in the case that it should be more than 800 X 1280 DP because user can see all information without any scrolling. We tested the performance and reliability of the system by running the application with 30 users simultaneously during one month to test fault tolerance, data accuracy, and response time. The result shows that the system can run continuously without any error and the average response time recorded for data transferring to control appliances is 0.635 seconds.

5. Conclusion

After the system was built, it was tested its performance, stability, and functionality whether it met the research's objectives or not. As mention before that one purpose of this system is to create a prototype that can assist the user in terms of energy consumption reduction and monitoring on Android devices, what we have found is that the system is easy to use because of smart devices can be touched on their screens. This creates a comfort to control each appliances or lights in houses. The ability to set a timer is very effective and useful in terms of solving the problem about energy overuse. It can be controlled by GPS in terms of calculate the distance between the device and the house and it can show the system status. The system status tells the user about how much they will pay at the end of each month and the cost of the electricity use will rise when the time goes by as they keep using appliances; this has shown that the user can monitor the price of electricity consumption in real-time. That means WebSocket protocol is suitable for this system in real-time data exchange.

In future works, the current sensor should be added to the system in terms of having a more accurate calculation of electricity use and cost. The Second suggestion, more Wi-Fi modules must be attached to each appliance to transfer data to Raspberry Pi because all appliances may not install in the same position and it is sometimes difficult to use cords for data transmission. All of these recommendations can function and cooperate with Raspberry PI and smart devices. The term "smart home" must come with the capability to automatically detect, calculate, and notify whatever it is programmed to do via Wi-Fi technology for providing its owners convenience and safety.

6. Acknowledgments

We would like to thank faculty of Science and Technology, Rajamangala University of Technology Thanyaburi for providing the infrastructure to do this research.

References

[1] Internet World Stats 2016 *Internet Usage Statistics* [Online]. Available: http://www.internetworldstats.com/stats.htm. [Accessed 20 January 2016].

[2] Internet World Stats 2016 *Mobile Internet - Mobile Phones* [Online]. Available: http://www.internetworldstats.com/mobile.htm. [Accessed June 2016].

[3] Ramlee R A, Othman M A, Leong M H, Ismail M M and Ranjit S S 2013 Smart home system using android application in *International Conference of Information and Communication Technology (ICoICT)* (Bandung).

[4] Kumar S 2014 Ubiquitous Smart Home System using Android Application *International Journal of Computer Networks & Communications* **vol. 6, no. 1** pp. 33-43.

[5] Pimentel V and Nickerson B G 2012 Communicating and Displaying Real-Time Data with WebSocket in *IEEE Internet Computing.*

[6] Vanessa W, Frank S and Peter M *WebSocket Frame Inspection with Google Chrome Developer Tools* (Apress).

[7] Berners-Lee T 1999 *HyperText Transfer Protocol* (World Wide Web Consortium).

[8] Karagiannis V, Chatzimisios P, Vazquez-Gallego F and Alonso-Zarate J 2015 A Survey on Application Layer Protocols for the Internet of Things *Transaction on IoT and Cloud Computing.*

[9] Kelly S T D, Suryadevara N K and Mukhopadhyay S C 2013 Towards the Implementation of IoT for Environmental Condition Monitoring in Homes *IEEE Sensors Journal* **vol. 13 no. 10** pp. 3846-53.

[10] Raspberry PI Foundation 2016 *Raspberry PI.* [Online]. Available: https://www.raspberrypi.org/. [Accessed February 2016].

The Hardware-Watchdog for Restart the Data-Logging Module because of Power-Drop in Street Lamps and Public Lamps Energy Saving Project

Kraimon Maneesilp[1] and Jaturapith Krohkaew[2]

[1, 2] Science and Technology Faculty Rajamangala University of Technology Thanyaburi Pathumthani, Thailand

Email : kraimon@rmutt.ac.th

Abstract. The data-logging module is a checking module using for saving energy of the street and public lamp. It must install in poor environments and difficult to access for maintenance. So, it should be designed to fault tolerant. The watchdog timer is one technique of the fault tolerant system. However, the power drop is the poor environment, which sometimes, the internal watchdog cannot solve it. This paper presents the hardware watchdog for solving this problem. It will be done hard reset on the controller board. It can solve the problem of electricity has gone down and improved the high fault tolerant to this system.

1. INTRODUCTION

The energy consumption management of street light and the public lamp is the interest of the public administration and local government. [1], [2] It saves the cost of welfare and safety of those living in the community. The governments can reduce this cost, but must meet the requirements of the standard of public lighting[3] or standard engineering practice[4] this can be done several techniques[5]-[7]. However, each method requires the data logging module for sending the values of luminance and energy consumption back to the server. Normally, the point of install this module is far away and difficult to access for maintenance. So, it should be designed to fault tolerant which some papers describe in developing such features [8]-[11] to keep the system running smoothly.

The system downtimes have many causes and prevent that from happening; it's going to be tough. One common is to increase the fault tolerance of the system is watchdog [12]-[13]. The watchdog is a timer to restart the systems. If the controllers unit cannot reset the watchdog before its timeout coming, the watchdog time will send the restart signal to all modules of controllers unit. By this technique, if controllers are a deadlock in some modules. After the restart, it can be fallout from this situation. However, from practical use finds some situation cannot fall out from halt by a restart from the watchdog. For example, the halt from electric power drops. The causes of watchdog cannot solve this case is the watchdog is soft restart. It sends the restart signal to all module with little time. It works if all module stays in the ready start condition. In the power drops case, some modules are not ready to start. For this case, the solution is to unplug and reconnected it to the power supply, called hard reset. That mean, the administrator will do it go to installation point and do it.

This article presents a solution to such by the simple digital circuit. This circuit composed of basic CMOS and simple electronic components. It is an external hardware watchdog that will be hard reset when the controller module does not send the reset signal to it before its timeout is coming. Because it can made the hard resets. Similarly to send the administrator go to the installation site and reset it, it can also reduce a lot of the maintenance costs and increase the fault tolerance into the systems.

2. HARDWARE-WATCHDOG DESIGN

The circuit is designed using standard CMOS. It uses IC 4060; 14-stage ripple carry binary counters to generate the timeout signal. In [14] the inside of 4060 is composed of the RC oscillator and 14 stage ripple counter. Its internal diagram shows in Figure 1.

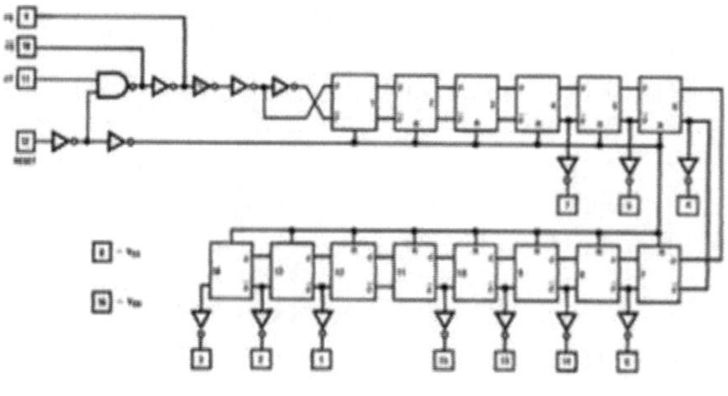

Figure 1. The internal diagram of IC 4060

It is evident that the clock signals from RC Oscillator will be sent to divided by 16, after that the signal will be sent to pin 7 and divided by two every pin before sent to next pin. The next pin of 7 is 5, 4, 6, 14, 13 and 15 respectively. After the signal passes the pin 15, it sent to divided by 4 and sent to pin 1 and divided by two every pin before sent to pin 2 and pin 3. The number's dividing of the clocks on each pin is represented by Table 1.

TABLE 1.THE NUMBER'S DIVIDING OF THE CLOCKS OF 4060

Pin	7	5	4	6	14	13	15	1	2	3
number's dividing	16	32	64	128	256	512	1024	4096	8192	16384

This circuit uses only pins 6, 14, 13 and 15 for user choice to set up. That mean the watchdog have four timeout choice: 128, 256, 512 and 1024 times of clock. The hardware watchdog circuit is shown in Figure 2. The circuit is composed of three sub-unit. First is the timer/counter unit; second is the relay unit and last is reset the unit. The timer/counter unit is shown in Figure 3.

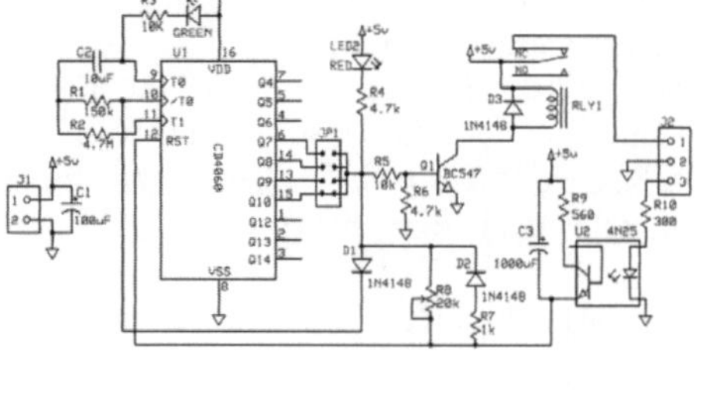

Figure 2. The circuit of hardware watchdog

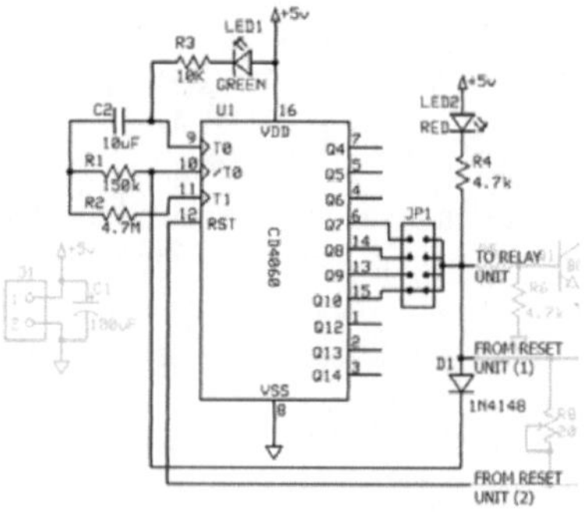

Figure 3. The timer/counter unit

This unit is set up the 4060 to the timer/counter. It is using internal RC oscillator to generate the timer clock. R1 and C2 configure the clock frequency that follows by (1).

$$Clock\ frequency = \frac{1}{2.2 \times R_1 \times C_2} \tag{1}$$

The cycle time of clock is

$$Cycle\ time = 2.2 \times R_1 \times C_2 \tag{2}$$

and the on period is the half of Cycle time then,

$$On\ period = 1.1 \times R_1 \times C_2 \tag{3}$$

From values of R1 and C2, the on period of the clock is 1.65 sec. So that, the on period of pin 6, 14, 13 and 15 is 3.5, 7, 14 and 28 minutes respectively. The on period is the timeout of watchdog when to select in this pin. The user can be select pin by setting the jumper JP1. When the timer runs over timeout, the output to relay unit will be change to a logic high and sent pass the diode D1 to pause the RC oscillator. The clock pause since the reset signal by C3 discharge in reset unit will send reset signal to 4060. The LED1 is a clock heartbeat, and LED2 is an indicator to show this time is not timeout overrun. The second unit is relay unit. It shows in Figure 4.

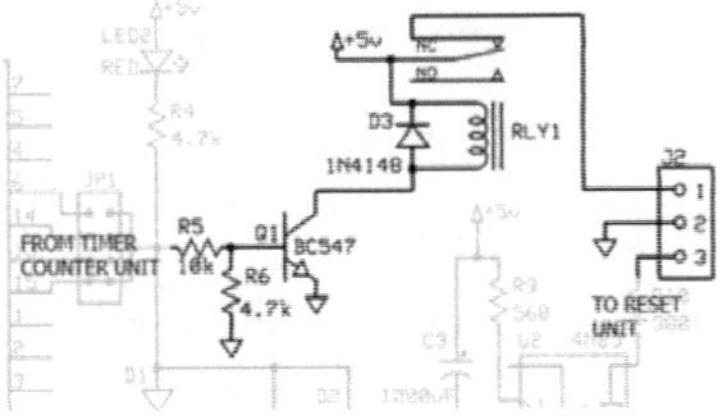

Figure 4. The relay unit

The relay unit receives the timeout signal from the timer/counter unit. This signal will drive the relay RLY1 to open and close the power supply to the controller board. If timeout signal is logic low, the transistor Q1 will be cut off, then the relay contactor of RLY1 is staying at Normal Close (NC). So, the controller board will connect to power supply and can operation.

When the timeout signal is logic high, the Q1 will be saturated, the relay contact of RLY1 will move to Normal Open (NO). In this time, the power supply will be cut from controller board and reconnect to Normal Close after the timeout signal is low. That mean the hard reset happens.

The last unit is reset unit, that show in Figure 5

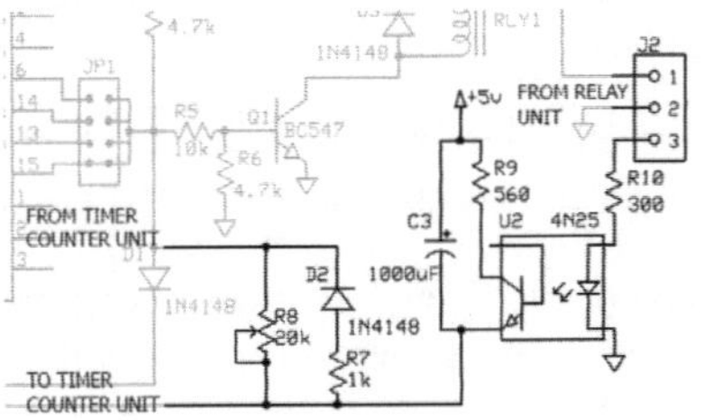

Figure 5. The reset unit

IOP Conf. Series: Materials Science and Engineering **185** (2017) 012020 doi:10.1088/1757-899X/185/1/012020

The reset unit operates on three times, first: when system power ON, second: after receiving the reset signal from the controller board and last is the timeout signal is coming. When system power ON, C3 will changing with R7 and D2, that made little pulse send to the reset (pin 12) of 4060, for clear all state in 4060. After receiving the reset signal from controller board at pin 3 of J2, the LED of isolator U2 will be lighting. When the phototransistor of U2 detects the light from LED, it will saturate then sends high level to the reset of 4060 and send the high-level pass R7 and D2 to pause the RC oscillator. The RC oscillator will be recovery after the reset signal from controller board return to low-level.

When the timeout signal is coming, the signal cannot pass R7 to the reset of 4060, because D2 is reverse bias. So the electric current will gently flow through R8 go to charging C3. The time to charge depends on the value of R8. When C3 is full, the high level of C3 will reset the 4060; that is restart timer again. So, the value of R8 is effective to off period of hard reset directly.

3. Implementation, testing, and result

The installation of watchdog board is showed in Figure 6. This board is inserted between the power supply and controller board and receive the watchdog reset signal from the controller board.

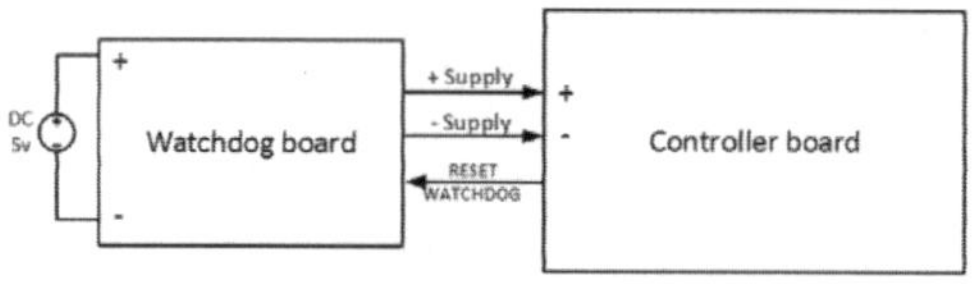

Figure 6. The install of watchdog board

The complete construct of watchdog board shown in Figure 7.

The experiments are assigned to two parts for check this solution. First: test running to timeout. This experiment for test the operates of watchdog that can be run to timeout and make the hard reset or do not. The last one is testing when the watchdog reset signal coming before timeout; the timeout will be a restart, and the output will continue to supply to controller board or do not. The experiments use a 5V power supply, one storage oscilloscope with two props and some copper wide. All equipment shows in Figure 8.

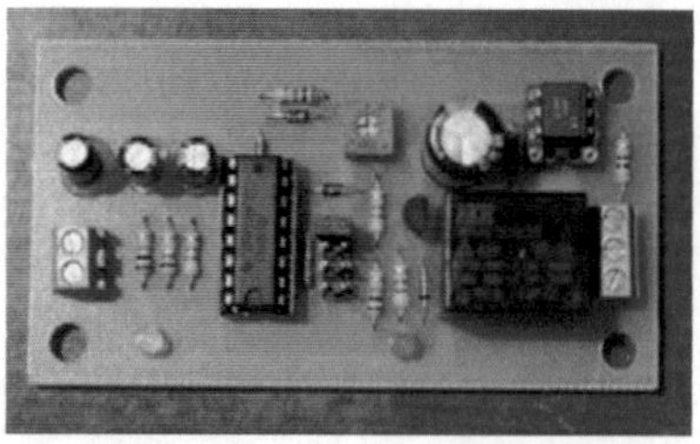

Figure 7. The complete watchdog board

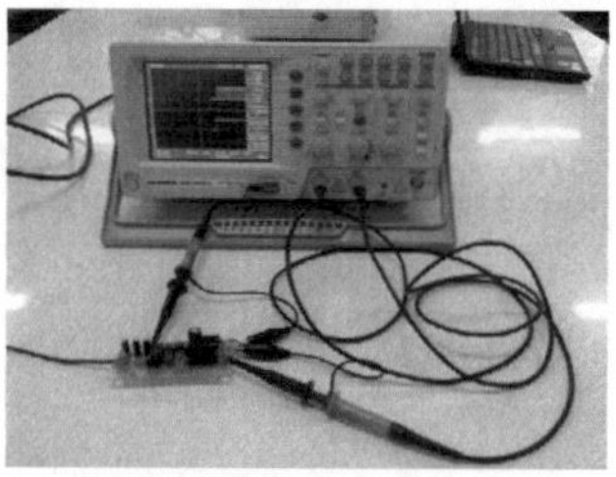

Figure 8. The equipment of experiments

The first experiment, will connect the power supply 5VDC to connector J1 and use the first channel of oscilloscopes to prop at pin 9 of 4060, that the output of RC oscillator and connect the second channel to pin 1 of J2. The result as shown in Figure 9. The first channel which props to oscillator shows the clock of the counter. The JP1 is jumping pin 6 to timeout-signal port; that mean the timeout is the clock divided by 128, or it has 128 pulse clock before the timeout is high level. When the timeout is high, the clock will pause, and the output is stopped supply to the controller board. After a few sec, the clock returns to continues and output return to supply, that is a complete hard reset. This result shows that the watchdog board can make the complete hard reset after time run over timeout.

The second experiment is setting the equipment similar the first experiment, but different the experimental procedure. After power on around 75 sec, we push the 5 sec pulse to pin 3 of J2 (reset-watch port) and repeat it every 75 sec. If the voltage on the second channel stays at 5v, that shows, the reset-watch signal can reset timeout of the circuit. The result shows in Figure 10.

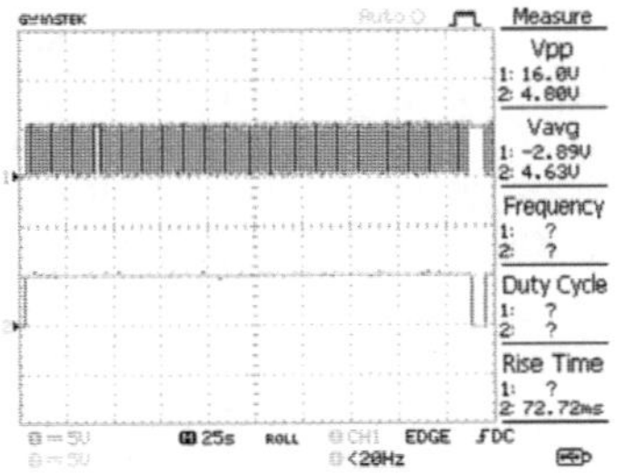

Figure 9. The result of experiment one

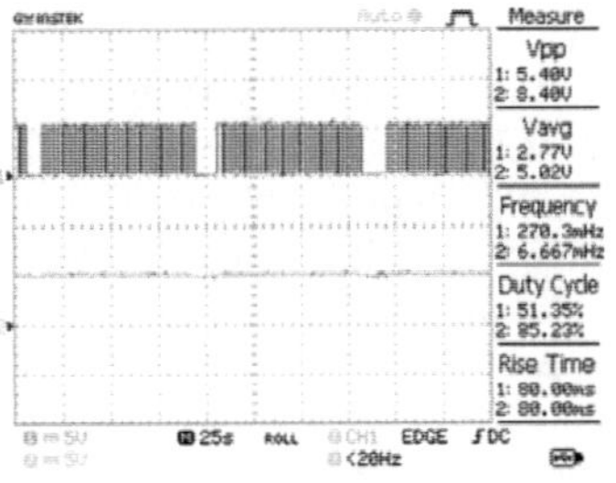

Figure 10. The result of experiment two

From the result, the voltage on the second channel can stay at 5 volts; that mean the controller board can get the power supply as long as the watchdog board has received the reset signal before its timeout.

We have implemented the hardware watchdog to the site1 of our project to restart the data logger and we got the result as shown in Figure 11. and Figure 12. They show that without the hardware watchdog our data logger can't be restarted since it was halted, so we lose the data of that period, however, after we implemented the hardware watchdog the data logger will be forced to restart after the watchdog counter is reach to zero and the restart time is too short, so we can get data continuously.

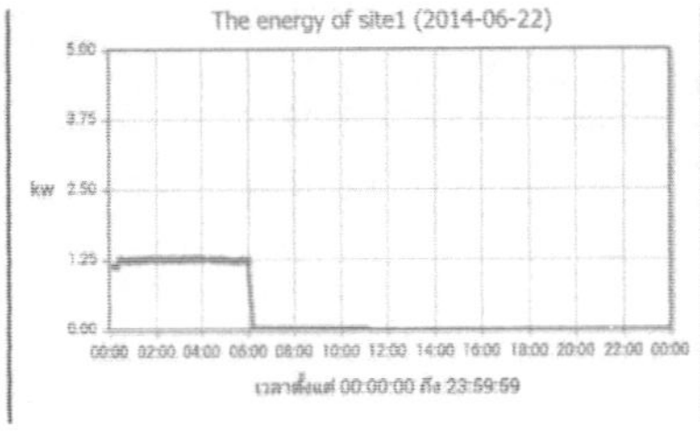

Figure 11. The graph of data lose before use watchdog

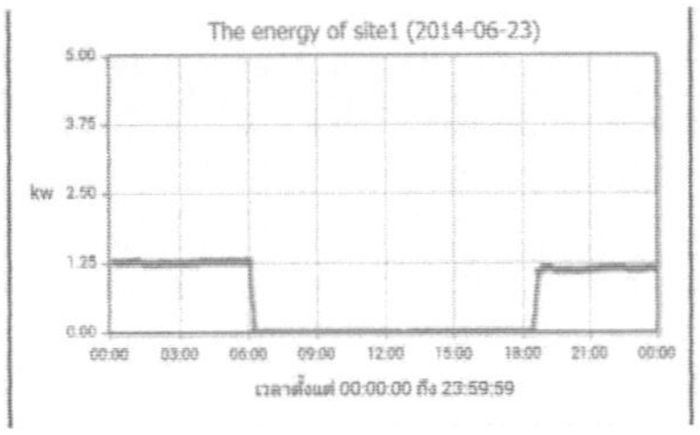

Figure 12. The graph of comploete data after use watchdog

4. Conclusion

The hardware watchdog was designed to solve the problem of restarting the system when it is halt by power drop with the internal watchdog. Other them, it can solve the critical problem that requires a hard reset. Because of its simplicity. It is resistant to errors. It is a good choice to increase the fault tolerance of the system.

Eventhough, the hardware watchdog can improve the high reliability of the system, but the real causes are never removed. So, it should be used in case of cannot solve the real causes. It should not use in case of not found the real causes of the problem.

Acknowledgment
We would like to thank you to the Provincial Electricity of Thailand for the financial support of the street lamps and public lamps energy saving monitoring project and RMUTT for the support of equipment's and lab.

IOP Conf. Series: Materials Science and Engineering **185** (2017) 012020 doi:10.1088/1757-899X/185/1/012020

References

[1] Karthikeyan M, Saravanan V and Vijayakumar S 2014 Cloud Based Automatic Street Light Monitoring System *Proceedings of 2014 International Conference on Green Computing Communication and Electrical Engineering (ICGCCEE 2014)* pp 1-6.

[2] Pattamavorakun S, Krohkaew J, Maneesilp K, Nilpruek P, Janseangrat P and Boonmee K 2015 Practical database system design production of administration management for energy economization of street and public lamps project *Proceedings of 2015 International Conference on Science and Technology (TICST 2015)* pp 377-80.

[3] American national standard institute 2014 *ANSI/RP-8-14 American National Standard for Roadway Lighting.*

[4] SaskPower 2013 *Standard Engineering Practice Section 4 Roadway Lighting Design Guide* June 2013.

[5] Shaligram A D and Bangali J A 2012 Energy Efficient Street Lightning control System Design for corridor Illumination *International Journal of Scientific &Engineering Research* **Vol. 3 Issue 4**.

[6] Riener A and Mullner R 2005 Energy efficient pedestrian aware Smart Street Lightning System *International Journal of Pervasive Computing and Communications* **Vol. 7 Issue 2** pp 147-61.

[7] Muñoz-Baca C A and Rivera-Mejía J 2016 Data Fusion for a Street Lighting Monitoring System Based on Statistical Inference and Fuzzy Logic *Proceedings of 2016 IEEE International Instrumentation and Measurement Technology Conference* pp 1-6.

[8] Mishra B C, Panda AS, Rout N K and Mohapatra S K 2015 A Novel efficient design of Intelligent Street Lighting monitoring system using ZigBee network of devices and sensors on Embedded Internet Technology *Proceedings of 2015 14th International Conference on Information Technology* pp 200-5.

[9] Patekar S R and Dongre M M 2015 An Intelligent System for Monitoring and Controlling of Street Light using GSM Technology *Proceedings of 2015 International Conference on Information Processing (ICIP 2015)* pp 604-9.

[10] Zanjani P N, Ghods V and Bahadori M 2012 Monitoring and Remote Sensing of The Street Lighting System using Computer Vision and Image Processing Techniques for The Purpose of Mechanized Blackouts *Proceedings of 2012 19th International Conference Mechatronics and Machine Vision in Practice (M2VIP)* pp 5-10.

[11] Zanjani P N, Bahadori M and Hashemi M 2013 Monitoring and Remote Sensing of The Street Lighting System using Computer Vision and Image Processing Techniques for The Purpose of Mechanized Blackouts (Development Phase) *Proceedings of 22nd International Conference on Electricity Distribution* pp 1-4.

[12] Abaffy J and Krajčovič T 2010 Software Support for Multiple Hardware Watchdog Timers in the Linux OS *Proceedings of 2010 International Conference on Applied Electronics (AE2010)* pp 1-3.

[13] Ďuríček M, Pohronská M and Krajčovič T 2010 Functional Prototype of Multiple Watchdog System Implemented in FPGA *Proceedings of 2010 International Conference on Applied Electronics (AE2010)* pp 75-8.

[14] National Semiconductor 1988 *CD4020BM/CD4020BC 14-Stage Ripple-Carry Binary Counters CD4040BM/CD4040BC 12-Stage Ripple-Carry Binary Counters CD4060BM/CD4060BC 14-Stage Ripple-Carry Binary Counters Datasheet* pp 1-6.

IOP Conf. Series: Materials Science and Engineering **185** (2017) 012021 doi:10.1088/1757-899X/185/1/012021

Development of a neuromorphic control system for a lightweight humanoid robot

Michele Folgheraiter, Amina Keldibek, Bauyrzhan Aubakir, Shyngys Salakchinov

School of Science and Technology, Robotics and Mechatronics Department, Nazarbayev University, Kazakhstan

E-mail: michele.folgheraiter@nu.edu.kz, b.aubakir@nu.edu.kz, akeldibek@nu.edu.kz, shyngys.salakchinov@nu.edu.kz

Giuseppina Gini, Alessio Mauro Franchi, Matteo Bana

DEIB, Politecnico di Milano, piazza L. da Vinci 32, Milano, Italy

E-mail: giuseppina.gini@polimi.it, alessiomauro.franchi@polimi.it, matteo.bana@mail.polimi.it

Abstract. A neuromorphic control system for a lightweight middle size humanoid biped robot built using 3D printing techniques is proposed. The control architecture consists of different modules capable to learn and autonomously reproduce complex periodic trajectories. Each module is represented by a chaotic Recurrent Neural Network (RNN) with a core of dynamic neurons randomly and sparsely connected with fixed synapses. A set of read-out units with adaptable synapses realize a linear combination of the neurons output in order to reproduce the target signals. Different experiments were conducted to find out the optimal initialization for the RNN's parameters. From simulation results, using normalized signals obtained from the robot model, it was proven that all the instances of the control module can learn and reproduce the target trajectories with an average RMS error of 1.63 and variance 0.74.

1. Introduction

For humanoid robots to be able to operate in household and public environments, they should be physically safe when sharing the workspace with humans and be autonomous and adaptable while performing locomotion, object recognition, and grasping.

The conventional approach towards ensuring safe human robot interaction is to prevent collisions, which is usually achieved by setting mechanical barriers, using special actuators, or internal force and torque sensors [1]. This cannot be implemented for humanoid robots as it leads to limited workspace, which is not acceptable in operating environments shared with humans where physical contact is inevitable. Therefore, the system should be designed such that forces applied on human are below the limit that can injure or inhibit the human operator. Our solution is to decrease the weight and the inertia of the links by employing light-weight materials and high power-to-weight ratio actuation systems. This has the effect of reducing the potential and kinetic energy of the robot limbs while operating and therefore, makes human robot interaction safer. Following these guidelines, the humanoid we designed has a total of 28 Degrees of Freedom

(DOF) and a weight of $3kg$. In particular, as depicted in figure 1, there are 6 DOF in each leg, 6 DOF in each arm, 2 DOF in the pelvis and 2 DOF in the neck. This will allow the robot to perform 3D dynamic walking and to reach different objects in its workspace by performing human-like motion.

Autonomy and the capability to adapt to new environments are very important features for a humanoid robot. The learning and execution phases occur at the same time in biological organisms as they are forced to exist and survive in a changing environment while simultaneously learning and improving their sensory-motor strategies. Therefore, a challenging goal for humanoid robotics is to develop control systems that can be executed and adapted while the robot is operating. By integrating Artificial Neural Networks (ANN) in the control architecture the robot can elaborate input signals in real time [2] and respond to stimuli through a set of coordinated motor actions [3], [4], [5]. RNN based on leak-integrate-and-fire neurons were proven to reproduce the behavior of microcircuits located in cortex of the human brain. In particular, they are capable to predict sensory inputs [6], compute tasks in real-time such as speech recognition and analysis of visual information [7] [8] [9] and generate reaching trajectories [10]. In [11] it was demonstrated that the presence in the circuit of feedback signals from read-out units with adaptable synapses allow the RNN to learn complex periodic signals without the excitation of additional inputs. The synapses can be adapted on the base of an error signal calculated as the difference between the target signal and the output of the neural circuit or by a reward-modulated Hebbian learning rule [12]. This represents a more biologically plausible mechanism that can be used in all the situations where the current performance of the robot depends on motor actions performed in the past [13].

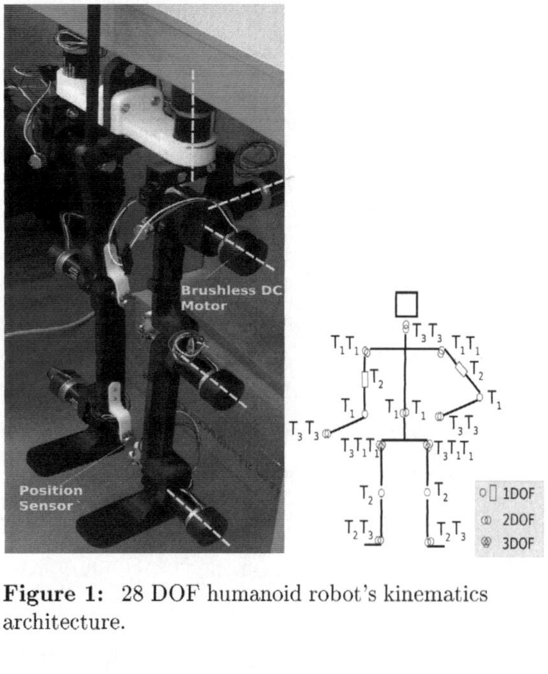

Figure 1: 28 DOF humanoid robot's kinematics architecture.

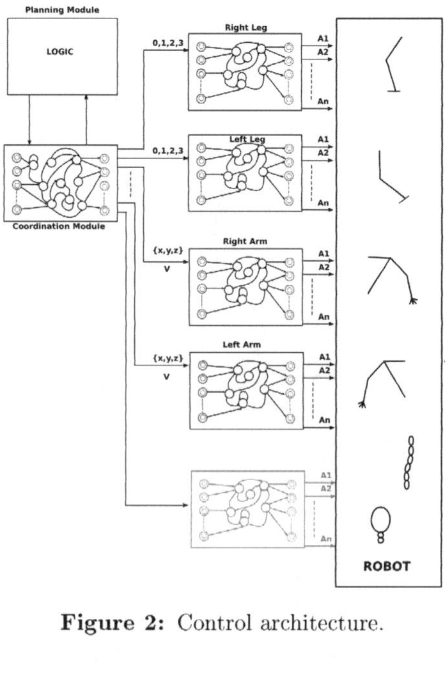

Figure 2: Control architecture.

The rest of the document is organized as follows: section 2 describes the control system and details a single RNN model, section 3 reports the preliminary results on a single control module capable to learn on training data obtained with a robotics simulator and to generate desired periodic joint trajectories. Finally, last section draws the conclusions and outlines future work.

2. Neuromorphic control architecture

The control architecture of the robot is organized in a hierarchical manner as depicted in figure 2. Neural modules that regulate the motion of each limb are at the lowest level. These modules have the same structure; in particular, each module is a RNN with fixed connections. The inputs and outputs of the module are routed towards an input and output layer of read-in and read-out units with tunable synapses. However, the number of read-in and read-out units can be adapted during the robot operation. This brings the advantage to introduce new inputs and outputs when required to perform a new task. Furthermore, if the read-in or read-out unit is wired to a specific piece of hardware (e.g. a joint motor or sensor) the architecture allows creating a new instance of the unit, adapting the correspondent synapses and passing smoothly the control of the hardware from previous to the new state. Low level modules that control leg motion have gait frequency, walking step size height as inputs whereas modules controlling arms receive target position, velocity, and instantaneous impedance as inputs.

Quantity	Value
Number of neurons N	100
Number of external inputs M	0
Number of read-out units L	5
Computational time, s	0.0001
Neuron time constant τ	0.013
Learning constant η	0.002
Decay learning constant dl	200
Chaos-modulation constant C	1.2

Table 1: The neural network parameters

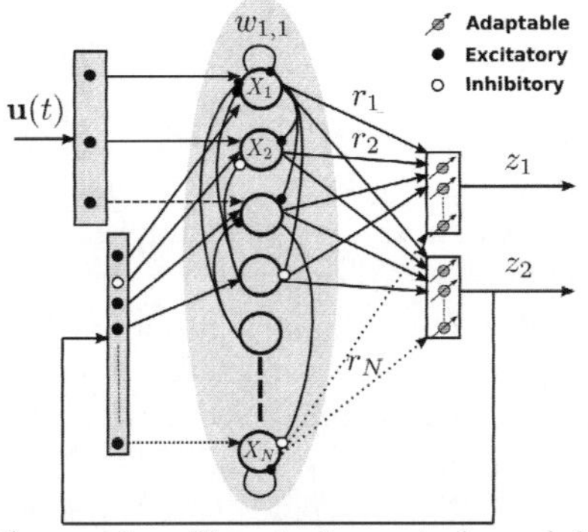

Figure 3: The single control module consists of a RNN where the outputs Z_i represent the read-out units.

At the highest level of architecture lies the coordination module that governs each neural module according to the specific tasks. This module receives commands from the planning module that implements the task; then, it sends feedback on the current state back. This layer of the control architecture is not based on RNN but rather on logic submodules.

2.1. Joints trajectory generation module

Each module of the control architecture depicted in figure 2 is capable to learn and generate different elementary motor paths that represent the building blocks to implement more complex locomotion and manipulation primitives.

The RNN we implemented (figure 3) consists of a reservoir of neurons which are randomly and sparsely connected through excitatory and inhibitory synapses of different strengths which are set randomly as well [10]. In particular, a single-layer architecture is used; i.e. it was demonstrated that addition of hidden layers does not provide any significant improvements [16]. Feedback loops that send the output of linear read-out units back to the network should be included to make learning possible without any external input. Results of experiments conducted by Sussillo and Abbott [14] showed that an initially chaotic network is faster to train and generates more accurate and robust output signals; therefore, the initial activity of our model is set to chaotic. In this RNN architecture only read-out unit weights are modified whereas other connections are not altered since their random initialization.

The RNN neuron is modeled by a differential equation (1), where x_i is the membrane potential, τ_i a time constant, C a parameter that modulates the chaoticity level of the RNN, r_j and u_j, the output and input of the jth neuron respectively, and z_j the signal generated by the read-out units. A nonlinearity (2) is added by a sigmoidal activation function that has also the function to limit the output of the neurons between $[-1\ 1]$.

$$\tau_i \dot{x}_i(t) = -x_i(t) + C\sum_{j=1}^{N} w_{ij}^{c} r_j(t) + \sum_{j=1}^{L} w_{ij}^{fb} z_j(t) + \sum_{j=1}^{M} w_{ij}^{in} u_i(t) \tag{1}$$

$$\Phi(x) = \frac{1 - e^{-kx}}{1 + e^{-kx}} \tag{2}$$

To explore new dynamic regimes [15] a Gaussian noise with zero average and variance equal to 0.005 is added to the neurons output and the read-out units (equations 3 and 4 respectively).

$$r_j(t) = \Phi(x_j(t)) + GNoise_j(t) \tag{3}$$

The probability of a connection between two neurons in the reservoir is $P = 0.1$ and the input and feedback weights are set randomly according to the uniform distribution $[-11]$. The reservoir comprises a total of 50% of inhibitory synapses and 50% of excitatory one. Although the circuit is able to generate periodic trajectories without additional signals, external inputs $\mathbf{u}(t)$ can be added to influence the neuron potentials according to the weights vector $\mathbf{W^{in}} \in \mathbb{R}^N$. Using this we can eliminate the presence of phase shift in the generated trajectories with time. Furthermore, an additional signal scaled by the L by N matrix $\mathbf{W^{fb}}$ forms a feedback from the outputs of read-out neurons.

The RNN's output $\mathbf{z}(t)$ is computed by a linear combination (4) of the outputs of the read-out units $\mathbf{r}(t)$ of a L by N matrix $\mathbf{W^{Ad}}$.

$$\mathbf{z}(t) = W^{Ad}\mathbf{r}(t) + GNoise(t) \tag{4}$$

The adaption of the matrix $\mathbf{W^{Ad}}$ is performed by a simple learning rule based on the error calculated as the difference between the target vector and the filtered version $\mathbf{z}(t)$ of the read-out unit vector $\bar{\mathbf{z}}(t)$ at the instant t as in (5).

$$\mathbf{Err(t)} = \bar{\mathbf{z}}(t) - F(\mathbf{z}(t)) \tag{5}$$

The read-out synapses are updated according to (6),

$$\mathbf{W^{Ad}}(t+1) = \mathbf{W^{Ad}}(t) - \eta(t)F(\mathbf{r}(t))\mathbf{Err(t)} \tag{6}$$

the learning constant decays according to the rule in (7).

$$\eta(t+1) = \frac{\eta(t)}{1 + t/dl}. \tag{7}$$

3. Simulation results

Training data was obtained from a simulation of robot static walking. We used the open source V-REP simulator [16] that has such functionalities as robot kinematics and dynamics, collision detection, virtual sensors and actuators, and particle dynamics. We acquired the position of the right leg joints during the execution of a static walk with a sampling time of $10ms$ for $220s$ and a total of 22000 samples.

The RNN we implemented is not computationally expensive [17] and can run on power-efficient computational units. It consists of 100 neurons and 5 read-out units receiving as inputs all the neurons outputs. The parameters of the circuit are reported in the table 1.

Particularly critical are the initialization of the weight matrix $W^C(i,j)$ for reservoir neurons, the chaos-modulation constant, the time constant, and the learning constant. The first three parameters regulate the RNN dynamics that transforms from an initially chaotic to a stable with the neuron outputs as periodic signals of constant amplitude. This allows to reproduce the target signals with a specific linear combination of the neurons outputs. The learning constant value should be higher at the beginning of the learning phase to force the RNN to reach a limit cycle (see figure 5) and then should be reduced.

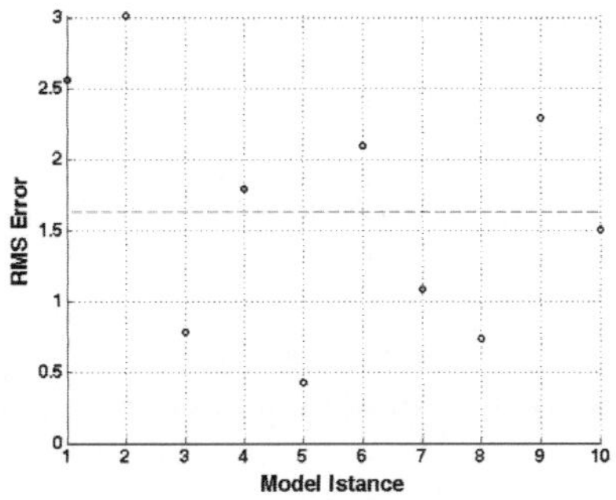

Figure 4: Overall RMS error for each of the 10 instances calculated on the validation set. The dashed line represents the mean of the RMS error.

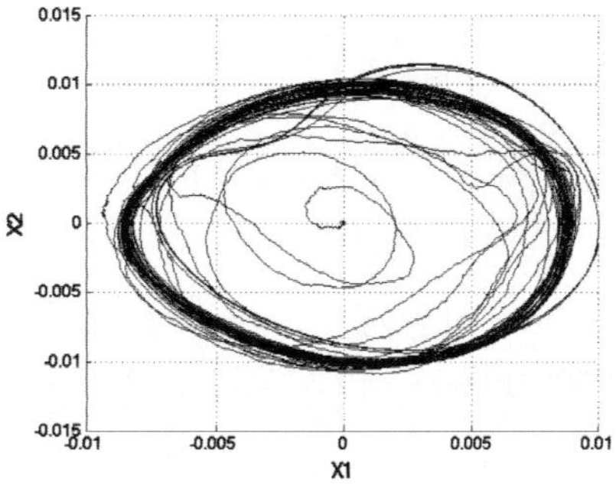

Figure 5: First two principal components of the neurons potential signals demonstrating the convergence ability of RNN circuit.

In total the adaptation phase lasts for $200s$. To find an optimal value of the matrix $W^C(i,j)$ we performed different experiments with different parameters kept constant and the matrix $W^C(i,j)$ randomly initialized for each model instance. The RNN was trained and the overall RMS error calculated as the sum of the RMS errors for each read-out unit. From figure 4 the average RMS error is 1.63 and variance is 0.74.

In figures 6 and 7 the outputs of the first two read-out unit are reported. The target trajectory is represented with a black dashed line and the RNN output with a continuous red line. From this measurements we can see that after the learning phase the RNN is able to reproduce periodic signals autonomously.

We also observed that after $100s$ the RNN outputs start to experience a phase shift. This can be eliminated by introducing an additional RNN input with a specific frequency that can serve as clock reference in the system.

4. Conclusion

In this paper we presented a lightweight humanoid robot equipped with a neuromorphic control architecture. The joints trajectories for a static walking are learned and reproduced by different modules organized into a hierarchical architecture. Each module is represented by a RNN consisting of 100 neurons modeled by a first-order differential equation. The neurons are coupled with fixed connections randomly initialized while special read-out units with adaptive synapses realize a linear combination of the neurons outputs in order to reproduce complex periodic signals. To optimize the weight matrix, ten experiments were performed where the training and validation phases were accomplished and the RMS error calculated. Preliminary simulation results show

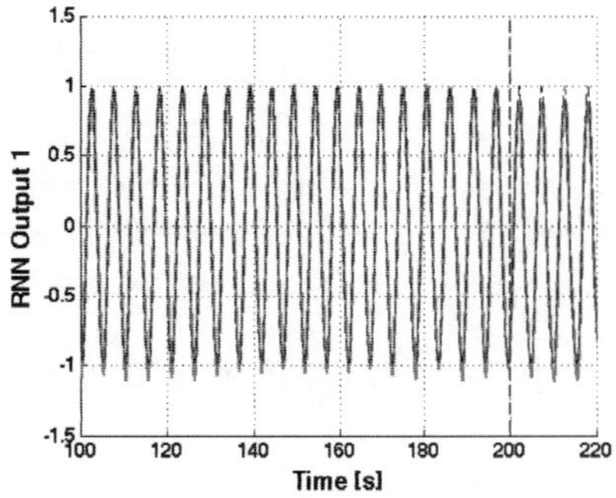

Figure 6: Read-out 1 reference signal with dashed line and RNN signal with red line.

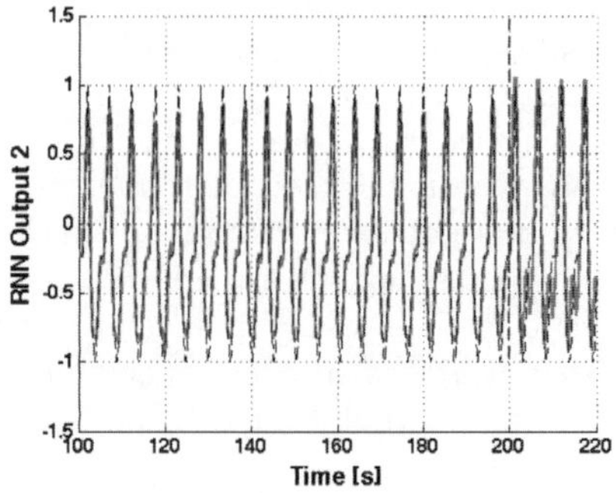

Figure 7: Read-out 2 reference signal with dashed line and RNN signal with red line..

that despite the small dimension of the network the module is able to reproduce the required trajectories. As a future work, we intend to implement the algorithm on a small computational unit e.g. a Raspberry Pi 3 or a BeagleBone and test the trajectories on the real prototype. It will be interesting also to integrate in the control architecture additional input signals from the sensory system of the robot. This will allow to produce more complex motor behaviors, e.g. alter the step size and height to avoid obstacles in its path.

Acknowledgment

This work was supported by the Ministry of Education and Science of the Republic of Kazakhstan under the grant and target funding scheme agreement #(348)226 – 2016.

Reference

[1] Lasota P A, Rossano G F and Shah J A *IEEE International Conf. on Automation Science and Engineering (Taipei, Taiwan, August 2014)* (IEEE) pp 339–344
[2] Folgheraiter M, Gini G, Nava A and Mottola N 2006 *Proc. of IEEE Robio06* (Kunming: IEEE) pp 1646–1651
[3] Chow T W S and Fang Y 1998 *IEEE Trans. on Industrial Electronics* **45** 151–161
[4] Carrillo R R, Ros E, Boucheny C and Coenen O J M 2008 *Biosystems* **94** 18–27
[5] Webb B and Consi T R (eds) 2001 *Biorobotics* (MIT Press)
[6] Markram H, Wang Y and Tsodyks M 1998 *Proc. Natl. Acad. Sci.*
[7] Legenstein R, Markram H and Maass W 2003 *Reviews in the Neurosciences* **14** 5–19
[8] Buonomano D V and Merzenich M M 1995 *Science* **267** 1028–1030
[9] Maass W, Natschlaeger T and Markram H 2002 *Neural Computation* **14** 2531–2560
[10] Joshi P and Maass W 2005 *Neural Computation* **17** 1715–1738
[11] Maass W, Joshi P and Sontag E D 2007 *PLoS Comput. Biol.* **3** e165
[12] Hoerzer G M, Legenstein R and Maass W 2014 *Cereb. Cortex* **24** 677–690
[13] Sutton R 1984 Ph.D. thesis University of Massachusetts Amherst
[14] Sussillo D and Abbott L F 2009 *Neuron* **63** 544–557
[15] Maass W 2014 *Proc. of the IEEE* **102** 860–880
[16] E Rohmer S P N Singh M F *Proc. of the International Conf. on Intelligent Robots and Systems (Tokyo, Japan, 2013)* (IEEE) pp 1321–1326
[17] Hoerzer G M, Legenstein R and Maass W 2012 *Cerebral Cortex* **24** 677–690

Bus Stops Location and Bus Route Planning Using Mean Shift Clustering and Ant Colony in West Jakarta

Kenny Supangat
System Information Dept., Multimedia Nusantara University, Banten, Indonesia
E-mail: kenny.supangat@student.umn.ac.id

Yustinus Eko Soelistio
System Information Dept., Multimedia Nusantara University, Banten, Indonesia
E-mail: yustinus.eko@umn.ac.id

Abstract. Traffic Jam has been a daily problem for people in Jakarta which is one of the busiest city in Indonesia up until now. Even though the official government has tried to reduce the impact of traffic issues by developing a new public transportation which takes up a lot of resources and time, it failed to diminish the problem. The actual concern to this problem actually lies in how people move between places in Jakarta where they always using their own vehicle like cars, and motorcycles that fill most of the street in Jakarta. Among much other public transportations that roams the street of Jakarta, Buses is believed to be an efficient transportation that can move many people at once. However, the location of the bus stop is now have moved to the middle of the main road, and its too far for the nearby residence to access to it. This paper proposes an optimal location of optimal bus stops in West Jakarta that is experimentally proven to have a maximal distance of 350 m. The optimal location is estimated by means of mean shift clustering method while the optimal routes are calculated using Ant Colony algorithm. The bus stops locations rate of error is 0.07% with overall route area of 32 km. Based on our experiments, we believe our proposed bus stop plan can be an interesting alternative to reduce traffic congestion in West Jakarta.

1. Introduction

According to the new start-stop index created by Castrol motor oil company [1], Jakarta has become number one city with the worst traffic in the world. The fact that Jakarta already has quite a variety of transportation like trains, angkot (public car established by private company), public bus (runs by a private company), ojek (a motorcycle runs by an individual person with low rate of fare), and the newest addition to it is the Busway (BRT-Bus Rapid Transit established by the Official Government of Jakarta) are still not proven to be an optimal solution for solving the traffic problem. Moreover, the actual issue that shows a huge contribution to the traffic problems was the increasing number of a private vehicle in Jakarta streets [2].The survey presented by the Statistic Department of Central Jakarta illustrate the number of a private two-wheel motorcycle is filling most of the roads with roughly 13.084.372 units or 74.67% on the road. In contrast, other vehicles especially buses only take 2.07% part of the population, while the other like private

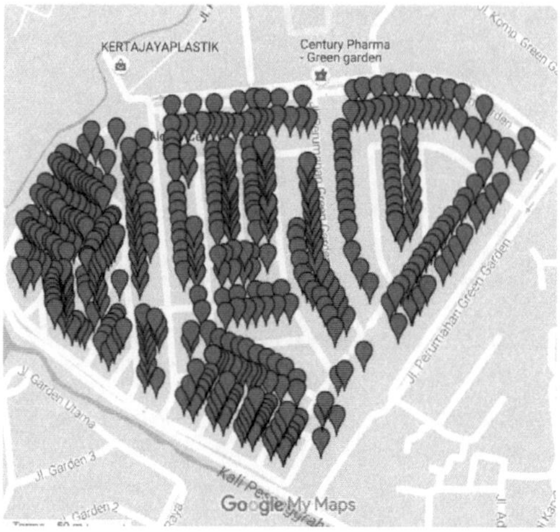

Figure 1: Example of part of West Jakarta on Google Maps. Red marks are annotated houses.

cars, trucks, buses, and governments vehicles are each takes the percentage of 18.64%, 3.84%, and 0.79% respectively [2].

The private vehicle owners mostly absorb the total usage of public transportation, especially the private car and motorcycle owner. The bus is one of a few public transportation that get the least attention from people due to some reason presented in the early survey. The survey asked 46 respondents (24 male and 22 female) about the transportation they usually use with their own reason. The survey shows that 63% (29 people) of the respondents don't mainly use bus because the bus is too far away from my home, the other reasons are there are no bus stop nearby my residence with 27%, and just 10% of the respondent felt awfully uncomfortable with the bus condition.

According to the data found during the survey, this research will be focusing on finding the optimal location for bus stops that will be located relatively near to the residential area, and the shortest route for the newly discovered bus stops location in this research.

2. Data Acquisition

The residential area located in West Jakarta holds the largest number of houses in Jakarta region [3]. Moreover, the region includes several well-known residential districts in the center of West Jakarta area (i.g. Green Garden, Kembangan, Kedoya district, and more), along with more than five thousand houses outside these districts. A large portion of West Jakarta have been chosen as the designated area for this experiment, one of the areas is shown in Figure 1 where the houses serve as the main data is marked as red pins in the figure.

The file containing the whole data will later serve as data input to finding the optimal bus stop locations by using mean shift algorithm implemented in python version 2.7.11.

3. Methodology

3.1. Bus Stop's Location Estimation

We apply mean shift [4-6] [11] (Equation 1) clustering algorithm to estimate the optimal location of the bus stops. The location of the bus stops are represented by the cluster center calculated by the mean shift, and the radius of the bus stop's service areas are the bandwidth of the clusters.

$$m(x) = \frac{\sum_{s\in S} K(s-x)s}{\sum_{s\in S} K(s-x)} \tag{1}$$

Where the difference of m(x) - x is called as mean shift in [4], x is initial centroid estimates (the longitude and latitude of inputs), S is the longitude and longitude of labeled houses, and K is the kernel used in the method (Gaussian kernel with $K(s-x) = e^{-c||s-x||^2}$). We also using a bandwidth or radius that represent how far is the service area of the new bus stop will be, in this case, we will set an arbitrary range of bandwidth of 500m as the default bandwidth for the first attempt in this experiment.

3.2. Bus Stops Routes Optimization

Previous studies have shown that ant colony can be used to solve optimal route problem (Bedi, 2007 [13]; Alves, 2010 [14]).The bus stops routes are optimized using ant colony algorithm [7] [8] [12] following:

$$P_{ij} = \frac{(\tau_{ij}^{\alpha})(\eta_{ij}^{\beta})}{\Sigma(\tau_{ij}^{\alpha})(\eta_{ij}^{\beta})} \tag{2}$$

where ,$P-ij$. is the probability of choosing the state, $T-ij$. is the intensity of the pheromone trail for each interstate, ,$N-ij$. is the visibility of a solution that would be selected by the ants. For the pheromone part, α is a parameter controlling the intensity of the pheromone trail where $\alpha \geq 0$, and β is a parameter controlling the visibility where $\beta \geq 0$ [10].

The new pheromone trail ($T-ij$) is recalculated in each state by:

$$\tau_{ij} = (1-\rho)\tau_{ij} + \rho\Delta\tau_{ij} \tag{3}$$

where ρ is the constant evaporation of pheromone trail $0 > \rho > 1$ [10].

The ant colony algorithm is implemented in 2013 version of Matlab. In our implementation we use $\alpha = 4$ and $\beta = 1$ as the main parameter of both the intensity of the pheromone trail and ants visibility. To calculate the pheromone trail, we set the $\rho = 0.15$.

4. Result and Analysis

4.1. Bus Stops Location

We estimate the location of bus stops using an arbitrary 500 meters bandwidth on Equation 1. It produces eight cluster centers that will be used as the new location for the new bus stop location (Figure 2a. Unfortunately, the result is not quite reliable, because in the real situation every road in the street is full of curves and not just a straight road.

To find the amount of error produced by the first result, a customized google maps API code is made to calculate how many houses miss the actual bus stop service radius or we can say further than 500m from house to bus stop. There are 594 houses or 7.72% out of 7962 total houses that miss the bus stop service radius, and also the other error information is the maximum, minimum, and average range of error of each house.

After seeing the error result, we decide to use various numbers of arbitrary bandwidth radius ranging from 450m, 400m, 350m, 300m, and 250m. From each of the result that can be seen in Table 1, and we choose one result that has the least error as the final solution.

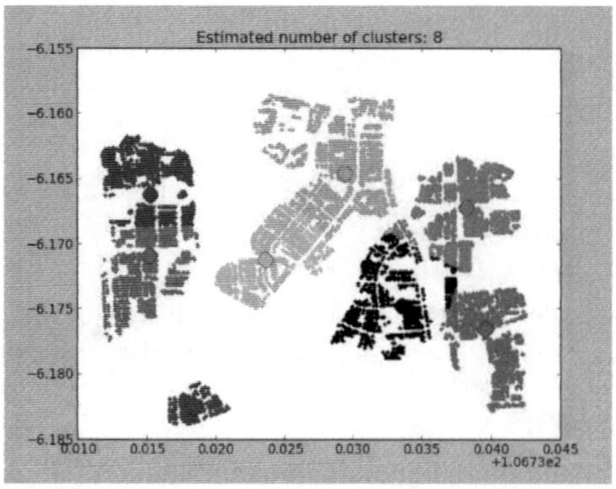

(a) Mean shift result with 500m bandwidth.

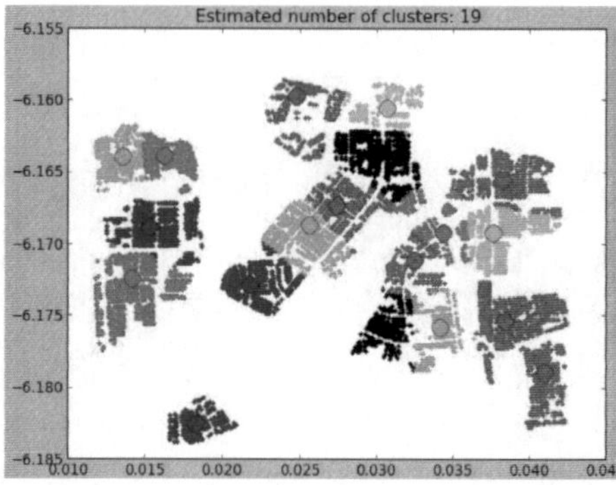

(b) Mean shift result with 350m bandwidth.

Figure 2: Mean shift results with two different bandwidths. Notice that smaller bandwidth produces more cluster center (bus stops).

Table 1: Summary of mean shift result. The best fit radius is 350 m indicated by the lowest error percentage.

Bandwidth	500 m	450 m	400 m	350 m	300 m	250 m
Total Error	594	22	57	6	10	7
Error Percentages	7.72%	0.28%	0.74%	0.07%	0.13%	0.09%
Max Error	0.4 km	0.07 km	0.08 km	0.09 km	0.11 km	0.1 km
Min Error	0.01 km	0.01 km	0.01 km	0.01 km	0.01 km	0.01 km
Median Error	0.09 km	0.03 km	0.02 km	0.02 km	0.03 km	0.07 km
Bus Stops Spawned	9	14	14	19	22	33

From the result illustrated by the Table 1, shows a various result from different bandwidth (radius). We pick the 350 m bandwidth because the result produces least error percentage with 0.07%, i.g. six houses miss the boundary of 350 m.

From the result using 350 m bandwidth in Figure 2b, we can see that there is 19 bus stop generated in the map, compare with Figure 2a that use 500 m bandwidth which only generates eight bus stops with higher bandwidth radius. This means the result of 350 m bandwidth can reach out smaller area than the result from 500 m bandwidth.

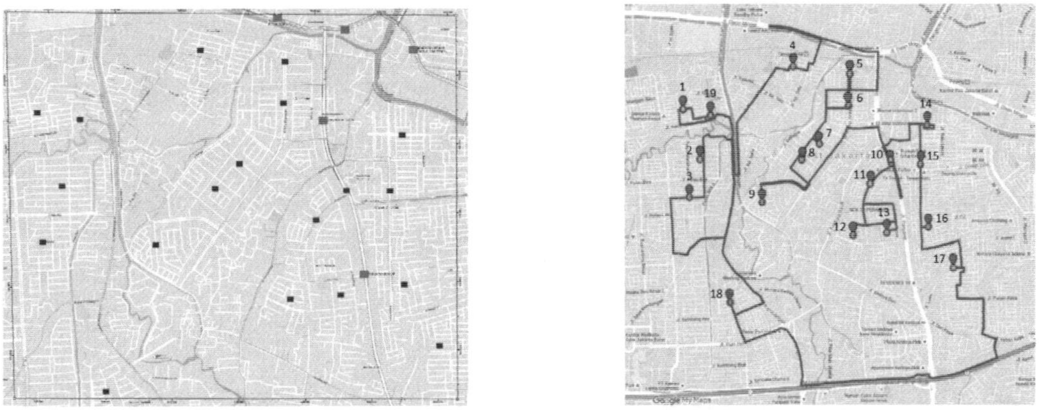

(a) The old and the new bus stops location. Red and blue squares represent the existing (old) and the proposed (new) bus stops location respectively.

(b) New proposed bus routes. Red marks are the bus stops. Blue paths indicate route traveled. The numbers show the sequence of the stops.

Figure 3: Mean shift results with two different bandwidths. Notice that smaller bandwidth produces more cluster center (bus stops).

The placement of the new bus stop and the existing bus stops can be seen in Figure 3a. The blue box represents the new bus stop, and the red box represents the old bus stop. We can see there are only 4 bus stops marked with a red box that currently operating and located in the middle of the main road far from the residential area . In the other hand, there are 19 new bus stops marked with a blue box scattered around the places so that many people can access the bus stop more easily across the residential area.

4.2. Bus Route

From the chosen result before, the geographical location of the new bus stop locations will be inserted into the source code as the input by using Matlab programming tools. There are several main inputs used in the like the number of ants, a number of cities (in here will be served as bus stop geographical location) and a number of iteration or repetition because the ant colony works in random and has many different kinds of possible outputs [9].

As presented in Figure 3b are the final result of ant colony method where the route leads from bus stop from number 1 through 19 and back to bus stop number 1. The visualization of the route is created using Google Maps feature.

The 26.44 km route comes from several run attempts of changing the parameter of ant colony method. From Table 2, 3, 4, and 5 are the attempt to find the shortest by changing some parameter in ant colony method which is number of ants, and number loop or attempt that the ants going to do to find the shortest path in that attempt.

5. Conclusion

This study has successfully determined the optimum solution for bus stop locations and its sub-optimal route. We prove that mean shift and ant colony algorithms can handle this particular problem, and should the case where the area is expanded further. And based on this experiments, we believe our proposed bus stop plan can be an interesting alternative to reduce traffic congestion in West Jakarta. Next we should expand the area of interest to cover wider Jakarta area and calculate more optimize bandwidth to find best fit radius.

References

[1] Toppa and Sabrina 2015(February) These Cities Have The Worst Traffc in the World, Says a New Index *TIME* Retrieved from http://time.com/3695068/worst-cities-tra_c-jams/?iid=sr-link1.

[2] Statistic Department Center of DKI Jakarta Province 2015 *Transportation Statistic of DKI Jakarta 2015* (DKI Jakarta BPS Province).

[3] StreetDirectory *Komplek Perumahan in Jakarta Barat* Retrieved from http://www.streetdirectory.co.id/indonesia/jakarta/landmark/zone/jakarta+barat/komplek+perumahan/.

[4] Fukunaga, Keinosuke, and Hostetler L D 1975 The estimation of the gradient of a density function, with applications in pattern recognition *IEEE Transactions on Information Theory* **21.1** 32-40.

[5] Cheng and Yizong 1995 Mean shift, mode seeking, and clustering *IEEE Transactions on Pattern Analysis and Machine Intelligence* **17.8** 790-9.

[6] Cheng, Yizong, and Fu K S 1985 Conceptual clustering in knowledge organization *IEEE Transactions on Pattern Analysis and Machine Intelligence* **5** 592-8.

[7] Dorigo, Marco, et al. 2008 eds. *Ant Colony Optimization and Swarm Intelligence: 6th International Conference, ANTS 2008, Brussels, Belgium, September 22-24, 2008, Proceedings* **Vol. 5217** (Springer).

[8] Dorigo, Marco, and Sttzle T 2009 Ant colony optimization: overview and recent advances *Techreport, IRIDIA* (Universite Libre de Bruxelles).

[9] Dorigo, Marco, and Gambardella L M 1997 Ant colonies for the travelling salesman problem *BioSystems* **43.2** 73-81.

[10] Haryanto, Wibowo A, Kusnadi A, and Soelistio Y E 2015 *Warehouse Layout Method Based on Ant Colony and Backtracking Algorithm* arXiv preprint arXiv:1508.04872.

[11] Comaniciu, Dorin, and Meer P 2002 Mean shift: A robust approach toward feature space analysis *IEEE Transactions on Pattern Analysis and Machine Intelligence* **24.5** 603-19.

[12] Dorigo, Marco, and Gambardella L M 1997 Ant colony system: a cooperative learning approach to the traveling salesman problem *IEEE Transactions on Evolutionary Computation* **1.1** 53-66.

[13] Bedi, Punam, et al. 2007 Avoiding trafic jam using ant colony optimization-a novel approach *International Conference on Computational Intelligence and Multimedia Applications* **Vol. 1** (IEEE).

[14] Alves, Diogo, et al. 2010 Ant colony optimization for traffic dispersion routing *13th International IEEE Conference on Intelligent Transportation Systems (ITSC)* (IEEE).

Using multiple linear regression model to estimate thunderstorm activity

W Suparta[1,*] **and W S Putro**[1]

[1] Space Science Centre (ANGKASA), Institute of Climate Change, Universiti Kebangsaan Malaysia, 43600 Bangi, Selangor Darul Ehsan, Malaysia.

Email: wayan@ukm.edu.my

Abstract. This paper is aimed to develop a numerical model with the use of a nonlinear model to estimate the thunderstorm activity. Meteorological data such as Pressure *(P),* Temperature *(T),* Relative Humidity *(H)*, cloud (*C*), Precipitable Water Vapor (PWV*)*, and precipitation on a daily basis were used in the proposed method. The model was constructed with six configurations of input and one target output. The output tested in this work is the thunderstorm event when one-year data is used. Results showed that the model works well in estimating thunderstorm activities with the maximum epoch reaching 1000 iterations and the percent error was found below 50%. The model also found that the thunderstorm activities in May and October are detected higher than the other months due to the inter-monsoon season.

Keywords: Multiple linear regression model, Meteorology data, Thunderstorm estimation

1. Introduction

Mesoscale convective systems (MCS) are large precipitation proportion in the convective system. MCS is an ensemble of thunderstorms with high precipitation index of 100 km or more in one direction [1]. The thunderstorm phenomena are categorized as the most deadly natural hazard in the world. Thunderstorm phenomena are electrical storms where cumulonimbus cloud produced lightning during heavy rain and thunder during precipitation. The estimation of thunderstorm frequency over MCS area is important to commercial space vehicle launch operation and power utilities. Suparta et al. [2] reported that the MCS activity in Tawau has high precipitation level that increased during the winter season (December, January, and February) and decreased during the summer season in June, July, and August. This showed that the MCS contributed to thunderstorm level in the Tawau area and hazardous for space activities.

Based on the impact of thunderstorm activity over Tawau area, this study aimed to estimate the thunderstorm frequency. Previous studies have shown that the simulation of thunderstorm activity is unable to evaluate the impact of a fault during an individual storm event [3, 4]. Based on these previous studies, it is suggested that Jacobi algorithm is applied in the multiple linear regression (MLR) to estimate thunderstorm activities. Six combination parameter inputs from meteorological parameter were constructed to develop a linear equation. From the meteorology data provided at each point of the Earth, the model can be easily implemented to other locations.

2. Methodology

2.1. Data and Location

Tawau is located in Sabah at East Malaysia (see Figure 1). The climate condition in this area is relatively hot and wet with average shade temperature about 26°C to 29°C at noon and falling to around 23°C at night. Looking the relationship between meteorology parameter and thunderstorm occurrence, the parameter inputs such as Pressure (*P*), Temperature (*T*), Relative Humidity (*H*), Cloud (C), Precipitable Water Vapor from radiosonde (RSPWV), and precipitation (Pr) were used to evaluate the variation of climate characteristic for thunderstorm activity. Data was taken over one year (1 January ~ 31 December 2012) for each parameter. The observation data were taken from Climate & Hydrological Section – Malaysian Meteorological Department (MetMalaysia) at TWU station. Moreover, RSPWV data and precipitation data were taken from the University of Wyoming website (*http://weather.uwyo.edu/upperair/sounding.html*) and the NASA website (*http://gdata1.sci.gsfc.nasa.gov*), respectively.

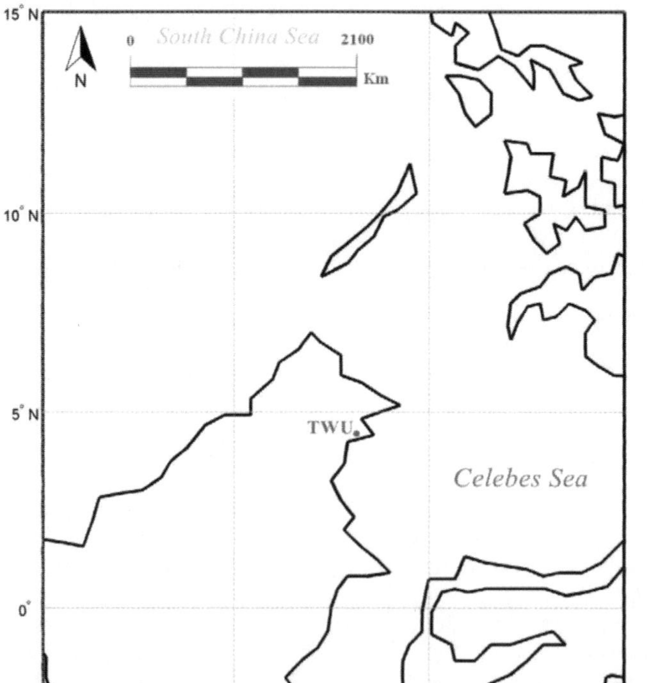

Figure 1. The location of TWU station for meteorological data observation

2.2. Data Processing

In this study, daily data were processed using MATLAB program to filter and enhance the quality of raw data. Figure 2 showed the flowchart of data processing for thunderstorm identification using meteorology parameters. The combinations of parameters were proposed to create a linear equation. The training data were processed to obtain estimation model with 1,000 iterations using the Jacobi algorithm. Five equation models were suggested to estimate the thunderstorm frequency based on the three seasons in 2012 which were summer, winter, and transition season (inter-monsoon). In this work, to design configuration input and output observation data using MLR method, the correlation and regression analysis to obtain R-square value between input (meteorological data, e.g. *P, T, H, C*, RSPWV, and Pr) and output (thunderstorm data) is performed.

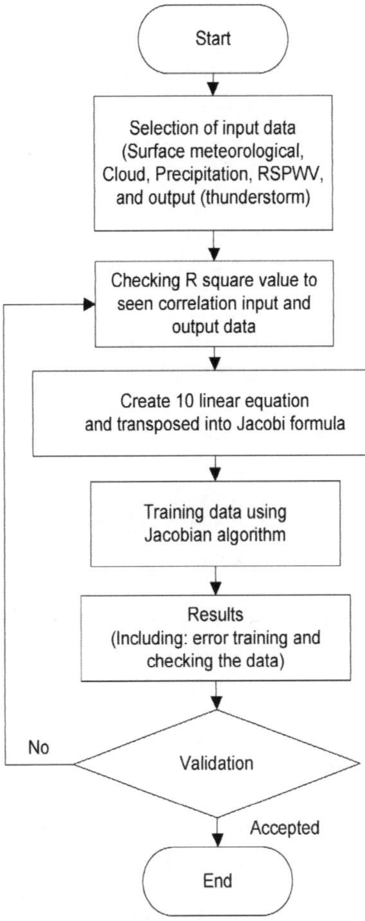

Figure 2. Flowchart of data processing and identification

2.3 Multiple Linear Regressions

The multiple linear regressions (MLR) equation can be described as an *algebraic equation* in which each term either a constant or the product of a constant and a single or more variable (linear equation). The multiple linear regressions are useful to characterize estimation model for any phenomena based on mathematics methods. In general, the MLR is written as:

$$Y_i = \beta_0 + \beta_1 X_{i1} + \beta_2 X_{i2} + \ldots + \beta_p X_{ip} + \varepsilon_i \quad (1)$$

where the least-squares estimates b_0, b_1, ... b_p are usually computed by statistical software, and the e_i are the residuals.

2.4 Jacobi Algorithm

The Jacobi method or Jacobi iterative method is an algorithm for determining the solutions of a diagonally dominant system of linear equations [5]. Each diagonal element in this method is solved accurately, and an approximate value is plugged in until the iteration process achieving convergent. The Jacobi method was used to find the best iteration by using a linear equation over six combination parameters. The Jacobi algorithm is expressed in Equation 2.

$$B = a_{x_{11}} + a_{x_{12}} + a_{x_{13}} + \ldots + a_{x_n} \quad (2)$$

where a_{xn} and B are not equal to zero for predictor and target parameters, respectively.

3. Results and Discussion

3.1. Meteorology Parameter

Figure 3 showed the variation of surface meteorology (*P, T,* and *H*), Cloud, RSPWV, and *Pr* from 1 January 2012 to 31 December 2012 as inputs the model. The variation of the first three of the parameter (*P, T,* and *H*) is observed in April, June and September reached the pressure to above 1010 mbar, June and August reached the temperature to above 28°C, January, February, and December reached the relative humidity to above 85%. Furthermore, the last three of parameter showed an extreme value of cloud density, RSPWV, and precipitation in June and July reached to above 7 oktas, February and December reached to above 80 mm, and February, May, and October reached to above 55 mm/day, respectively. On the other hand, winter I and winter II from Figure 3 is estimated from the month of January – February and December, respectively. The summer is defined from June to August and the other months are intermonsoon. The minimum precipitation and high temperature indicated that the clear day was obtained in the middle of July. In addition, the rainy day during the winter season occurred in January and February 2012and December.

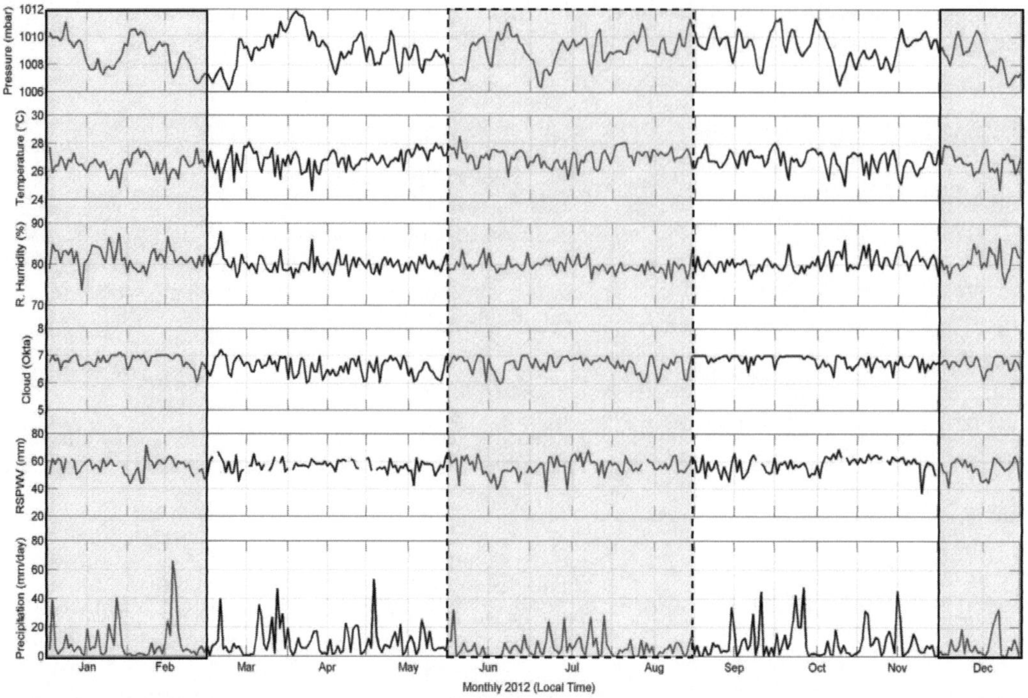

Figure 3. The variation of meteorology parameter over Tawau area. The shaded area shows the winter I, summer, and winter II, respectively.

3.2. Thunderstorm Estimation

A standard error regression (S) and R-squared (R-sq) values are examined to show the best relationship between input and output developed for estimation. Results show that the configuration input with relative humidity and RSPWV obtained a good result as compared to another configuration with S and R-sq are obtained below 50% and 6%, respectively. After the best configuration input was obtained, the input with *H* and RSPWV are proposed to develop a MLR equation by using the observation data in 2012. The selection configuration inputs were processed to find the maximum iteration. Table 1 shows the MLR equation for three seasons in 2012 that was processed using Jacobi method. The maximum iteration for x_1 ~ x_2 is presented in two - three steps with two configurations for each season. The Python software was used to obtain the estimation model. In this study, 1000 iteration (max) was used to find the optimum input model for estimation process.

Table 1. The estimation of *H* and RSPWV using Jacobi algorithm for the season in 2012

Season	The Jacobi Equation	Solution
Winter I	$3.48 = 0.0462\ x_1$	$x_1 = 75.324$
(JF) month	$3.33 = 0.0514\ x_1 - 0.0104\ x_2$	$x_2 = 52.08$
Summer	$-2.64 = -0.0294\ x_1$	$x_1 = 89.795$
(JJA) month	$-2.58 = -0.0294\ x_1 + 0.00092\ x_2$	$x_2 = 61.19$
Winter II	$3.49 = 0.0470\ x_1$	$x_1 = 74.255$
(Dec) month	$3.90 = 0.0465\ x_1 + 0.0080\ x_2$	$x_2 = 55.89$
Transition I	$1.60 = 0.0248\ x_1$	$x_1 = 64.516$
(MAM) month	$1.61 = 0.0264\ x_1 - 0.00217\ x_2$	$x_2 = 42.96$
Transition II	$2.12 = 0.0325\ x_1$	$x_1 = 65.230$
(SON) month	$2.06 = 0.0371\ x_1 - 0.00736\ x_2$	$x_2 = 48.92$

Note that x_1 and x_2 stand for *H* and RSPWV, respectively

The estimation of parameter *H* and RSPWV from two configuration equations for each season was obtained using Jacobi algorithm. However the negative result in x_2 (RSPWV) indicated that the value is out of range. Furthermore, the estimation result using MLR equation is presented in Figure 4. As can be seen on the figure, five equations were used in the Jacobi methods which were winter I, transition I, summer, transition II, and winter II. Based on the investigation, the minimum and maximum *RSPWV* and *precipitation* are 38 ~ 82 mm and 0 ~ 76 mm/day, respectively. The estimation result of thunderstorm frequency for each season was obtained by using rounding number technique. For example, if the output reached < 0.5 would mean that there is no event (value = 0) while output at > 0.5 indicate that an event may occurred (value = 1).

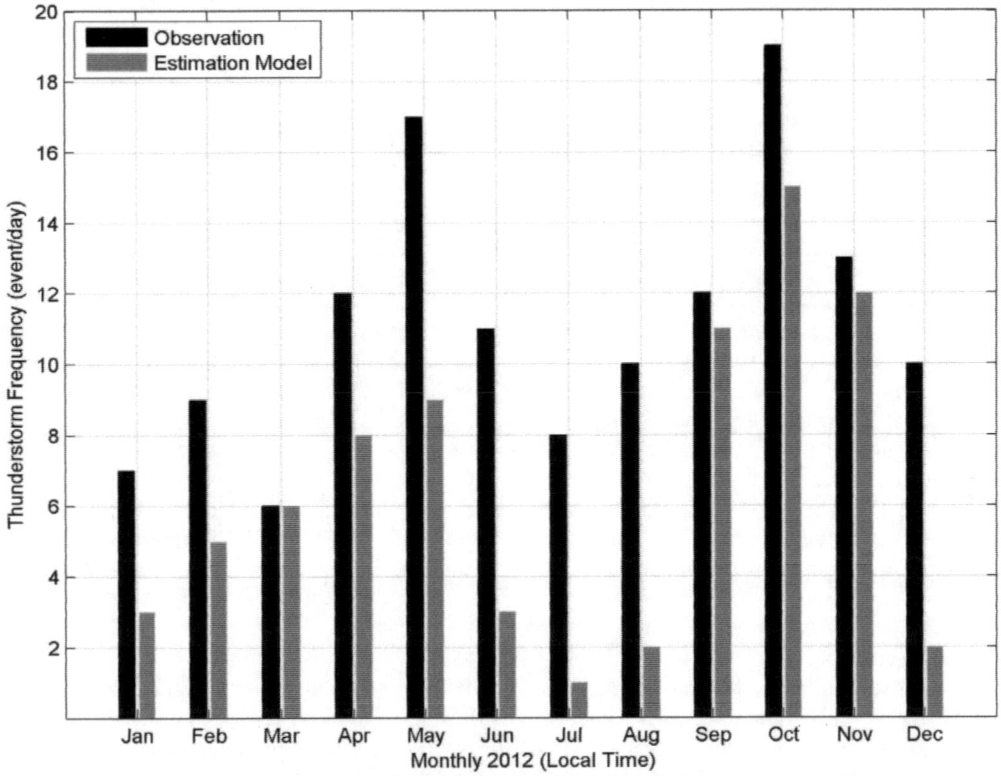

Figure 4. The estimation model of thunderstorm activity over TWU station

In order to evaluate the performance of the MLR model in estimating thunderstorm activity, the MLR model was compared with *Dvorak* technique and Adaptive Neural Fuzzy Inference System (*ANFIS*) with *Fuzzy Clustering Method* (*FCM*). Figure 5 shows the comparison between the three models. The model used 1,000 epoch and five layers (fixed layer ANFIS) and found with maximum error < 1%. From the figure, it can be seen that the three models comparison provides a strong relationship (> 0.75 at the 99% confidence level), where ANFIS FCM reached the highest correlation followed by the *Dvorak* technique and the Numerical model. In addition, all three methods reached minimum error (see Table 2) as indicated by root mean square error (RMSE), mean absolute error (MAE), and percent error (PE). ANFIS FCM provides good results followed by the *Dvorak* technique and the MLR model. From the comparison analysis, the PE for MLR is worse than the other models which possibly due MLR cannot fully capture the linearization for 1,000 iterations.

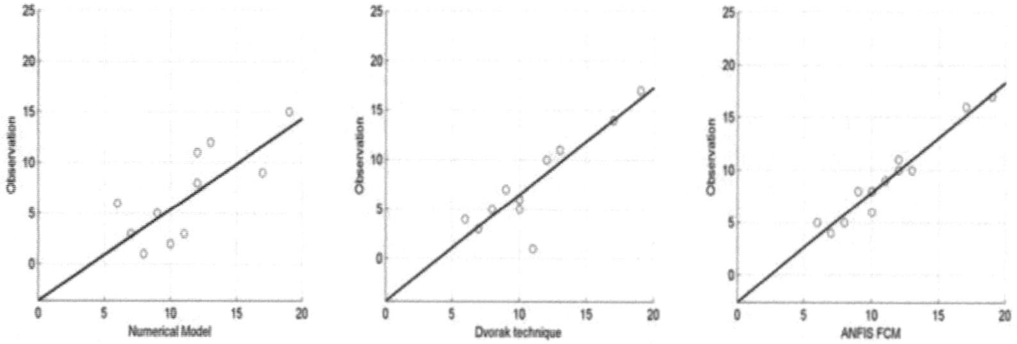

Figure 5. Comparison of observation data between (a) Numerical Modeling, (b) *Dvorak* technique, and (c) ANFIS FCM model

Table 2. The comparison analysis of thunderstorm activity in the TWU station in 2012

Method	Correlation (R^2) value	RMSE	MAE (%)	PE (%)
MLR Model	0.8911	4.7697	4.4167	39.7288
Dvorak technique	0.9352	3.8944	3.6667	36.4322
ANFIS FCM	0.9812	1.8930	1.7500	17.2097

4. Conclusion and Future Work

A MLR model using Jacobi method was successfully carried out to estimate thunderstorm activity over Tawau Area for the year of 2012. The model was developed based on the two configuration inputs (*H* and RSPWV) and one output of thunderstorm. The thunderstorm activity in Tawau was found highest in the inter-monsoon season in March ~ May (winter to summer) and September ~ November (summer to winter), respectively. More than 12 events per month were obtained in this season indicating that the region is very convective. The comparison between MLR model, *Dvorak* technique, and ANFIS FCM found that the MLR model have percent error almost 40%, which is not suitable for estimation of complex thunderstorm. Finally, the ANFIS model is more advantage in constructing the estimation of thunderstorm activity, and therefore, it is suggested as a prediction model in the next studies.

Acknowledgements

This study is funded by the Ministry of Science, Technology and Innovation (MOSTI) under Science Fund 01-01-02-SF1100 grant and UKM Publications Rewards: GP-K017155. We would like to thank the Malaysian Meteorological Department (MetMalaysia) for providing the meteorological data, NASA website for providing the precipitation data, and the Department of Atmospheric Science, University of Wyoming for providing the RSPWV data.

References

[1] Houze R A Jr 2004 Mesoscale Convective Systems *Reviews of Geophysics* **42** 8755-1209
[2] Suparta W, Putro W S, Singh M S J, Asillam M F 2015 Characterization of GPS and Meteorological Parameters for Mesoscale Convective Systems Modelover Tawau, Malaysia *Advanced Science Letters* **21**(2) 203-206
[3] Albar A, AL-Khalaf A K and Abdel-Basset H 2015 Radar Rainfall Estimation of a Severe Thunderstorm over Jeddah *Atmospheric and Climate Sciences* **5** 302-316
[4] Balijepalli N, Venkata S, Richter Jr C, Christie R and Longo V 2005 Distribution system reliability assessment due to lightning storms *IEEE Trans. Power Delivery* **20** 2153-2159
[5] Yang X I A and Mittal R, 2014 Acceleration of the Jacobi iterative method by factors exceeding 100 using scheduled relaxation *Journal of Computational Physics* **274** 695–708

IOP Conf. Series: Materials Science and Engineering **185** (2017) 012024 doi:10.1088/1757-899X/185/1/012024

Sugarcane Land Classification with Satellite Imagery using Logistic Regression Model

F Henry[1], D E Herwindiati[2], S Mulyono[3], and J Hendryli[4]

[1,2,4] Faculty of Information Technology, Tarumanagara University, Letjen S. Parman No. 1, Jakarta, Indonesia

[3] The Indonesian Agency for the Assessment and Application of Technology,, Kawasan Puspitek, Serpong, Tangerang, Indonesia

E-mail: 535120072@fti.untar.ac.id, dyahh@fti.untar.ac.id, sidik.mulyono@bppt.go.id, jansonh@fti.untar.ac.id

Abstract. This paper discusses the classification of sugarcane plantation area from Landsat-8 satellite imagery. The classification process uses binary logistic regression method with time series data of normalized difference vegetation index as input. The process is divided into two steps: training and classification. The purpose of training step is to identify the best parameter of the regression model using gradient descent algorithm. The best fit of the model can be utilized to classify sugarcane and non-sugarcane area. The experiment shows high accuracy and successfully maps the sugarcane plantation area which obtained best result of Cohen's Kappa value 0.7833 (strong) with 89.167% accuracy.

1. Introduction

Indonesia is one of the countries in the world with great potential in the agricultural sector, which is important in the development of the country's economy. Crops plantation will have a high economic value if managed well. Monitoring agriculture process is one of the ways to manage crops plantation that can be done using satellite imagery data [1]. Utilizing satellite data has many benefits, such as obtaining information on the area that is difficult to reach like forests, swamps, and mountains which can help in agricultural planning such as irrigation [2], growth phase monitoring [3], and crop yield estimation [4]. Remote sensing technology also makes it easier and cheaper to map a very large area.

Sugarcane is one of the basic materials in sugar production and the increase in demand in Indonesia results in a greater need for suitable land for the plantation. Indonesia has a large sugarcane plantation area with more than 1.3 million farmers families involved and 2.5 to 3.0 million metric tons annual sugar production potential [5]. This paper discusses the mapping of sugarcane plantation from Landsat-8 satellite imagery. The process is broken down into two steps: training and classification. Polygon data, surveyed by the Indonesian Agency for the Assessment and Application of Technology (BPPT), are used as input for the training stage. The reflectance of sugarcane is selected according to the known wavelengths of sugarcane vegetation to calculate the normalized difference vegetation index (NDVI) using the NIR and red band reflectance values. After obtaining the set of data, multivariate analysis is used to build the logistic regression model to detect the sugarcane plantation in another area.

2. Related Works

As early as 1991, Lee-Lovick and Kirchner [6] had shown that sugarcane canopies give important spectral information which can help the task of classification from satellite imagery. The relationship between spectral reflectance and sugarcane canopy architecture had also been reported by [7]. As [8] reported, the spectral response of sugarcane is also influenced by chlorophyll, carotene, and other various pigments. Therefore, several vegetation indices, such as normalized difference vegetation index (NDVI) [9] and enhanced vegetation index (EVI) [10], are usually utilized as features to recognize the sugarcane plantation. A study from [11] shows a high relationship between MODIS NDVI and rainfall patterns on the sugarcane plantation.

Numerous models had been proposed in previous researches. Unsupervised classifier showed promising result of detecting sugarcane plantation from MODIS satellite imagery and EVI features in [10]. Another study from [12] used k-means clustering to classify sugarcane area from EVI temporal data after noise reduction using wavelet analysis. Lee-Lovick and Kirchner [6] studied the classification from Landsat TM imagery and found that band 1, 2, and 3 spectral reflectances are useful for identification of sugarcane. A study from [13] showed a model to discriminate sugarcane from coffee and citrus plantation in Brazil with a high accuracy of 95%.

3. Logistic Regression

Logistic regression is a mathematical model approach that can be used to describe the relationship of several independent variables on a dichotomous dependent variable [14]. In this paper, the model parameter (x) is the time series of NDVI value and the class variable (y) is state of sugarcane class (1) or non sugarcane class (0). in the case of sugarcane land classification. The logistic regression model predicts the class through the probability calculated from the sigmoid function $g(x)$.

$$h_\theta(x) = g(\theta^T x) = \frac{1}{1 + e^{-\theta^T x}} \tag{1}$$

where the range value is $0 \le h_\theta(x) \le 1$. The target y can determine from the value of h_θ as in Equation 2.

$$y = \begin{cases} 1 & h_\theta(x) > 0.5 \\ 0 & h_\theta(x) \le 0.5 \end{cases} \tag{2}$$

The regression coefficient θ is estimated using maximum likelihood estimation (MLE). The MLE of logistic regression cannot be expressed in a closed-form, so an iterative method is used to maximized the likelihood function. We use gradient descent algorithm to minimize the cost function $J(\theta)$.

$$J(\theta) = -\frac{1}{m} \sum_{i=1}^{m} y^{(i)} \log h_\theta(x^{(i)}) + (1 - y^{(i)}) \log \left(1 - h_\theta(x^{(i)})\right) \tag{3}$$

where m is the number of training samples, $y^{(i)}$ is the target, and $x^{(i)}$ are the features of each data in the training set. The gradient of the cost function is defined as

$$\frac{\partial}{\partial \theta_j} J(\theta) = \frac{1}{m} \sum_{i=1}^{m} \left(h_\theta(x^{(i)}) - y^{(i)}\right) x_j^{(i)} \tag{4}$$

and at each iteration in the gradient descent updates the parameter $\theta_{new} = \theta_{old} - \alpha \frac{\partial}{\partial \theta_j} J(\theta)$ where α is the learning rate.

The completed algorithm of the training step is as follows:

(i) Choose the initial value for θ, learning rate α, and error rate ϵ.

(ii) Repeat the following steps until the error converges $|J(\theta_{new}) - J(\theta_{old})| > \epsilon$.

(a) Calculate the new sigmoid function h_θ and $\frac{\partial}{\partial\theta_j}$.

(b) Simultaneously update the parameter θ.

(iii) The parameter θ after the iterations is the coefficient of logistic regression model.

4. Experiments and Results

4.1. Data

The input to the model is the series of normalized difference vegetation index (NDVI) for sugarcane and non-sugarcane land, consisting of 17 periods in a year. These periodic data follow the one-year phenology of sugarcane. The spectral values are obtained using ENVI 5.1 software.

The NDVI is a numerical indicator that uses a visible band and near-infrared of the electromagnetic spectrum and is widely adopted in the analysis of green vegetation in remote sensing. The formulation of NDVI is described as

$$\text{NDVI} = \frac{NIR - RED}{NIR + RED} \tag{5}$$

where NIR is the value of near-infrared spectral reflectance and RED is the value of visible spectral reflectance. The NDVI-series will be formatted and combined sequentially by the date of capture through the layer stacking process and served as the input in the training step.

The aim of the training step is to find the best logistic regression model to classify the sugarcane plantation. It will predict the sugarcane and non-sugarcane area through the logistic probability model, calculated from the sigmoid function in Equation 1 and 2. Meanwhile, the testing process is done to test the accuracy of the sugarcane area classification using logistic regression. We have 1800 locations of sugarcane and non-sugarcane area, which are employed in these two experiments. The first scenario utilizes 60% of it as the training data and the second scenario uses 25%, as the rest are for testing the trained model. These locations are from 17 satellite images of Kawadenan, East Java from May 2014 through May 2015.

4.2. Experimental Settings

The preprocessing of the Landsat-8 satellite imagery data is largely done using the ENVI software. For the classification process, we develop classification software using .NET framework 4.5. Our initial logistic regression parameters are initialized to be 0 and the learning rate is set to 0.5. We run the model training for 800 iterations. In every hundredth iteration, we save the parameters and calculate the performance.

4.3. Evaluation Measure

We evaluate the performance of the logistic regression model for sugarcane plantation area classification by calculating the prediction accuracy, Cohen's kappa statistic, and pseudo-R^2. Cohen's kappa statistic, first introduced by [15] can be interpreted as a interclass correlation coefficient [16]. It can also be used to measure the interrater reliability [17]. Cohen's kappa coefficient κ can be defined as follows:

$$\kappa = \frac{OA - AC}{1 - AC} \tag{6}$$

where OA is the sum of true positive and true negative divided by the total number of data and AC can be calculated as in Equation 7.

Table 1. The interpretation of the Cohen's kappa coefficient

Value of κ	Strength of Agreement
< 0.20	Poor
$0.21 - 0.40$	Fair
$0.41 - 0.60$	Moderate
$0.61 - 0.80$	Good
> 0.80	Very Good

$$AC = \left(\left(\frac{P}{n}\right) * \left(\frac{P'}{n}\right)\right) + \left(\left(\frac{N}{n}\right) * \left(\frac{N'}{n}\right)\right) \tag{7}$$

where P is the sum of true positive and false positive, N is the sum of false negative and true negative, and P' is the sum of true positive and false negative. The Cohen's kappa coefficient shows the strength of agreement between two variables as in Table 1 [18].

To evaluate the goodness-of-fit of the logistic model, some pseudo-R^2 measures have been developed, such as Efron's [19] and McFadden's [20] pseudo-R^2. These are "pseudo" measures because they look like R^2 in a sense that they are on the same scale. The Efron's pseudo-R^2 can be calculated as in Equation 8 with y_i is the target and $\bar{y}$ is the mean of the target data.

$$\rho^2 = \frac{\sum_{i=1}^{n} \left(y_i - h_\theta(x_{(i)})\right)^2}{\sum_{i=1}^{n} \left(y_i - \bar{y}\right)^2} \tag{8}$$

Meanwhile, McFadden's pseudo-R^2 is defined as in Equation 9 where $LL(B)$ is log-likelihood of the full model and $LL(0)$ is log-likelihood of the null model. While the range of ρ^2 is similar to R^2 in regular regression analysis, yet the measurement cannot be determined with a standard index of R^2 and standard of "good fit" [20].

$$\rho^2 = 1 - \frac{LL(B)}{LL(0)} \tag{9}$$

4.4. Results

Figure 1 shows the pseudo-R^2 for each iteration in the training stage. The figure shows that the best model is acquired in 500 iterations. Although this is the case, we found that in the testing stage performance of the model from 600 iterations gives the best prediction accuracy. Table 2 shows the best Cohen's kappa coefficient and accuracy from the model.

It is interesting to see that with only 25% data as the training input, the trained logistic model can identify the sugarcane plantation area with more than 85% accuracy. The Cohen's kappa coefficient also shows that the model is good. Confusion matrix in Table 3 further shows the performance of this model on 25% proportion of the data.

In Figure 2 we show the mapping of sugarcane plantation in Kawadenan region, East Java. On the left is the ground truth data obtained by the BPPT and on the right is the classification result from the logistic regression model. The green area shows the sugarcane plantation area detected by the model, while the red area is the sugarcane real plantation. The logistic regression model detected much of the plantation area, but some misclassifications are observed.

5. Conclusion and Future Works

We have presented the logistic regression model for classifying sugarcane plantation area in Kawadenan region, East Java. The parameters learned in 500 iterations show the best pseudo-R^2 but in the testing stage, the prediction accuracy is lower than the parameters obtained from

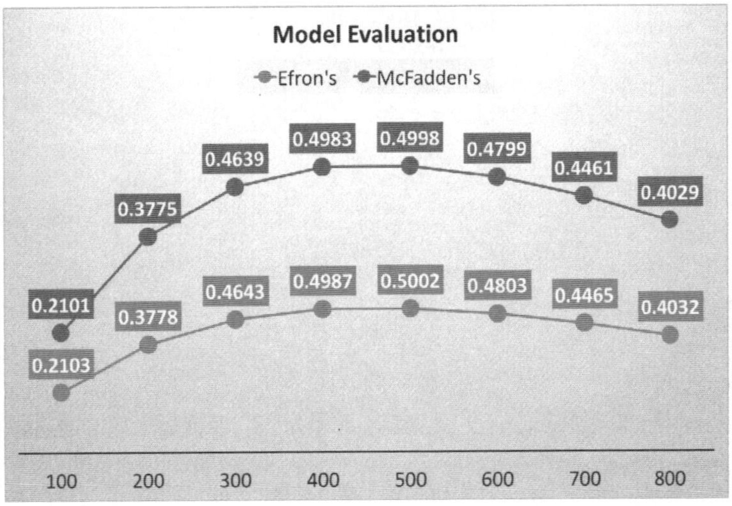

Figure 1. Pseudo-R^2 from our model training

Table 2. Cohen's kappa coefficient and accuracy from our experiments

		Proportion	
		60% training	**25% training**
Cohen's Kappa	Training	0.6 (good)	0.6933 (good)
	Testing	0.7833 (good)	0.7141 (good)
Accuracy	Training	80%	84.667%
	Testing	89.167%	85.704%

Table 3. Confusion matrix from the model trained only with 25% proportion of the data

		Prediction	
		Sugarcane	Non-sugarcane
Actual	Sugarcane	542	60
	Non-sugarcane	133	615

600 iterations. The model from 600 iterations predicted the sugarcane plantation with more than 85% accuracy even with only 25% of 1800 ground truth data. For the future works, we plan to utilize better learning algorithm, such as stochastic gradient descent to help to minimize the cost function of logistic regression faster and more effective. More features to describe the sugarcane plantation can also be explored further to increase the prediction accuracy.

References

[1] Frolking S, Qiu J, Boles S, Xiao X, Liu J, Zhuang Y, Li C and Qin X 2002 *Global Biogeochemical Cycles* **16**
[2] Kim Y, Evans R G and Iversen W M 2008 *IEEE Trans. on Instrumentation and Measurement* **57** 1379–1387
[3] Moran M S, Inoue Y and Barnes E 1997 *Remote sensing of Environment* **61** 319–346
[4] Shanahan J F, Schepers J S, Francis D D, Varvel G E, Wilhelm W W, Tringe J M, Schlemmer M R and Major D J 2001 *Agronomy Journal* **93** 583–589
[5] Wright T and Meylinah S 2016 Indonesia sugar annual report 2016 Tech. rep. U.S. Department of Agriculture Foreign Agricultural Service
[6] Lee-Lovick G and Kirchner L 1991 *Proc. the Australian Society of Sugar Cane Technology* vol 13 pp 124–129

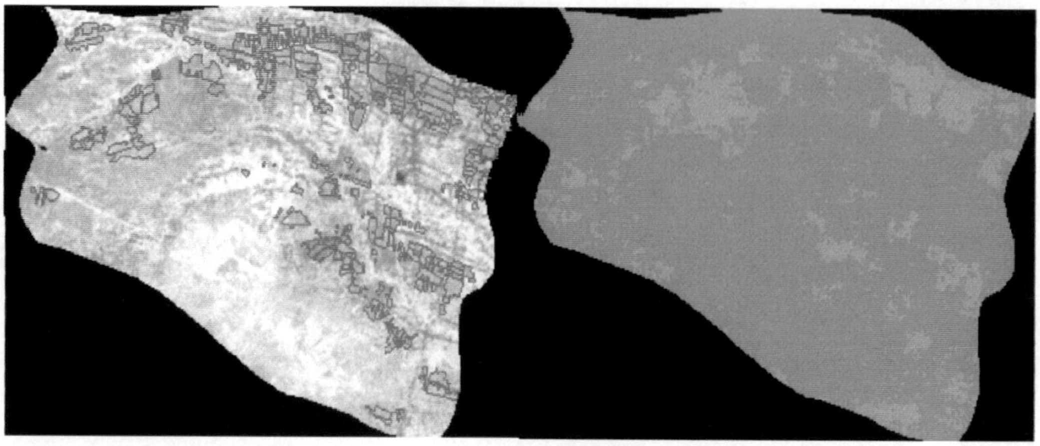

Figure 2. Mapping of ground truth sugarcane plantation area (left) and logistic regression model prediction (right)

[7] Galvao L S, Formaggio A R and Tisot D A 2005 *Remote Sensing of Environment* **94** 523–534
[8] van der Meer F, De Jong S and Bakker W 2002 *Imaging spectrometry* (Springer) pp 17–61
[9] Rahman M R, Islam A and Rahman M A 2004 *Plan Plus* **1** 1–12
[10] Xavier A C, Rudorff B F, Shimabukuro Y E, Berka L M S and Moreira M A 2006 *Int. J. Remote Sensing* **27** 755–768
[11] Gunnula W, Kosittrakun M, Righetti T L, Weerathaworn P and Prabpan M 2011 *Australian J. Crop Science* **5** 1845
[12] Rudorff B, Adami M, Aguiar D A, Gusso A, Silva W F and De Freitas R M 2009 *IGARSS (4)* pp 252–255
[13] Tardin A, Deassuncao G and Soares J 1992 *Pesquisa Agropecuaria Brasileira* **27** 1355–1361
[14] David K and Mitchel K 2010 *Logistic regression: A self learning text* 3rd ed (New York: Springer–Verlag Inc)
[15] Kohen J 1960 *Educ Psychol Meas* **20** 37–46
[16] Ebel R L 1951 *Psychometrika* **16** 407–424
[17] Smeeton N C 1985 *Biometrics* **41** 795
[18] Azen R and Walker C M 2011 *Categorical data analysis for the behavioral and social sciences* (Routledge)
[19] Scott Long J 1997 *Advanced quantitative techniques in the social sciences* **7**
[20] McFadden D, Talvitie A, Cosslett S, Hasan I, Johnson M, Reid F and Train K 1977 *Urban Travel Demand Forecasting Project, Phase* **1**

Colour Based Image Processing Method for Recognizing Ribbed Smoked Sheet Grade

Ike Fibriani, Sumardi, Alfredo Bayu Satriya, and Satryo Budi Utomo
Department of Electrical Engineering, University of Jember, Jalan Kalimantan 37 Kampus Tegalboto, Jember, Indonesia

E-mail: ikefibriani.teknik@unej.ac.id, smardi10@gmail.com, alfredobayu.teknik@unej.ac.id, satryo.budiutomo@yahoo.com

Abstract. This research proposes a colour based image processing technique to recognize the Ribbed Smoked Sheet (RSS) grade so that the RSS sorting process can be faster and more accurate than the traditional one. The RSS sheet image captured by the camera is transformed into grayscale image to simplify the recognition of rust and mould on the RSS sheet. Then the grayscale image is transformed into binary image using threshold value which is obtained from the RSS 1 reference colour. The grade recognition is determined by counting the white pixel percentage. The result shows that the system has 88% of accuracy. Most faults exist on RSS 2 recognition. This is due to the illumination distribution which is not equal over the RSS image.

1. Introduction

Rubber is a commodity which is used as a material in many appliances of human life. There are two kinds of rubber, namely natural rubber and synthetic rubber. Natural rubber is made from latex of para rubber tree, while synthetic rubber is made from crude oil. These two kinds of rubber have different utilizations, although they actually able to replace each other. The natural rubber is often used when the crude oil supply is in a few amount and vice versa. Indonesia is the second largest natural rubber producing country in the world. In 2011, Indonesia contributed 27.3 % of rubber production. The manufactured natural rubber in Indonesia generally is formed into rubber crepe or rubber sheet. The natural rubber which is formed into sheet is called Ribbed Smoked Sheet (RSS).

There are some processes in forming natural rubber into RSS, namely harvesting rubber from the tree (the rubber is in the form of latex), latex dilution, latex coagulation which forms the latex into rubber, rubber milling, rubber smoking process, and rubber sorting process. Sorting process is necessary because previous process produces different qualities/ grades of rubber sheet which also have different economy aspect in the market, they are RSS 1, RSS 2, RSS 3, RSS 4 and RSS 5 [?]. In traditional sorting process, RSS is sorted manually by the worker based on the colour of the RSS. This sorting process has some weaknesses because it depends on human perception which is not consistent. Due to this problem, an automatic system which can recognize the RSS grade is needed. The system must consist of a programmed camera which will capture the RSS surface and use it to automatically recognize the grade of the RSS with the use of an image processing method on its algorithm.

A previous work [?] has designed RSS grading system which has a good result with nearly 81% precision, but the system proposed is quite complex which needs 10 seconds of access time and still has error in recognizing the RSS quality. RSS grade can be recognized based on its colour, therefore a system based on image processing is suitable for RSS automatic sorting system process. Image processing has been used widely in many aspect including agriculture industry. Paper [?] used colour based image processing for fruit sorting system, another paper [?] used it to check the rice grade. Moreover, colour based image processing has been used in estimating air quality and fire detection [?, ?, ?]. This article proposes a based image processing method for recognizing the RSS so that the sorting process becomes more accurate and faster than the traditional one.

2. Ribbed Smoke Sheet

RSS is divided into six categories based on The Green Book-International Standard for Quality and Packing for Natural Rubber Grades, although many manufacturers do not always produce all the six categories. In this article, only three categories (RSS 1, RSS 2 and RSS 3) will be discussed and observed. These three RSS can be distinguished by its physical properties. RSS 1 must be dry, clean, strong, sound, and free from blemishes, resinous matter (rust), blisters, sand, dirty packing and any other foreign matter. RSS 2 must have at most 5% of rust and dry mould, while RSS must have at most 10% of rust and dry mould [?]. The samples of these three RSS rubbers are shown on Figure ??.

Figure 1. Comparison of different grades of RSS surface from left to right RSS 1, RSS 2, and RSS 3 (http://www.psc-rubber.com/)

3. Method and Model

The complete RSS recognizing algorithm is shown in Figure ??. RSS 1 surface has no rust and mould so it is considered having a smooth surface and a brighter colour than other RSS. So, the idea of the algorithm is making the pixel colour of RSS 1 as a reference point to detect the percentage of how much rust and mould on an unknown grade RSS which will be sorted. The percentage of the rust and mould of the RSS will determine its grade. When the RSS is captured by the camera, its image is on the RGB model. This model of image causes some difficulties when we try to compare between the RSS 1 reference pixel with an unknown grade RSS since a pixel in the image consist of three values namely red, green and blue. For simplifying the comparison, both RSS 1 reference pixel and unknown grade RSS are transformed into grayscale image. Grayscale image is chosen because its pixel only has one value and it represents the darkness level or the luminance of the RSS which can distinguish a smooth surface and a surface with rust and mould. RGB image is transformed into grayscale image using Equation ??.

$$I_{GS}(x,y) = \frac{I_R(x,y) + I_G(x,y) + I_B(x,y)}{3} \quad (1)$$

where $I_{GS}(x,y)$ is a grayscale value of an image pixel, $I_R(x,y)$ is red value of a RGB image pixel, $I_G(x,y)$ is green value of a RGB image pixel, $I_B(x,y)$ is blue value of a RGB image pixel and (x,y) is the pixel coordinate in an image.

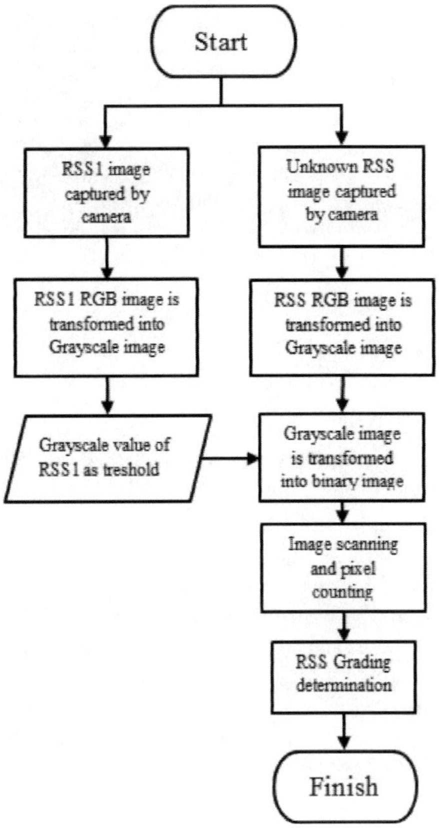

Figure 2. Flowchart of colour based image processing algorithm for recognizing RSS grade

After grayscale transformation, the unknown grade RSS is transformed into binary image using Equation **??** with the grayscale reference pixel as the threshold. This process produces a binary image with white pixel represents the smooth surface and black pixel represents the rust and mould. Next step, the binary image is scanned to count the percentage of the black pixel (rust and mould) compare to all pixels in the image. Having the percentage of black pixel, the unknown grade RSS can not be directly graded using the The Green Book rule (RSS 1 has 0%, RSS 2 has 5%, and RSS 3 has 10% of black pixels/rust and mould) because the RSS image captured by the camera is always affected by the environment illumination which often produces noise or black pixel on the binary transformation. In order to find the right percentage of white pixel for grading rule, the statistical data must be obtained. Twenty sheets of each RSS 1, RSS 2, and RSS 3 are tested to find the white pixel percentage. The process ends when the RSS grade has been recognized. Each of the colour model transformation is shown on Figure **??**.The system which has been designed is tested to recognized the grade of a hundred sheets of known grade RSS.

$$I_{bin}(x,y) = \begin{cases} 0 & if \quad I_{GS}(x,y) < T \\ 255 & if \quad I_{GS}(x,y) \geq T \end{cases} \tag{2}$$

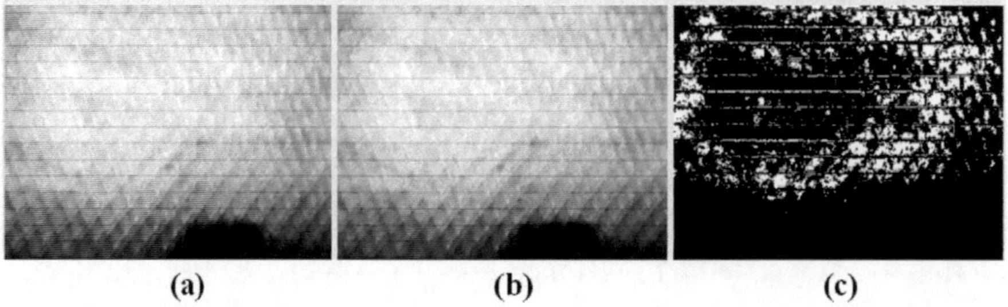

Figure 3. A sample of RSS image in (a) RGB model captured by camera, (b) Grayscale model after transformation, (c) Binary model after threshold process

where $I_{bin}(x, y)$ is a binary value of an image pixel, $I_{GS}(x, y)$ is a grayscale value of an image pixel, (x, y) is the pixel coordinate in an image and T is the threshold value.

4. Result and Discussion

Statistical result of pixel percentage in each grade of RSS is shown in Table ??. It has been discussed on previous chapter that the white pixel percentage which represents smooth surface on RSS 1 can not reach 100% (0% rust and mould) due to illumination effect. It makes the RSS 1 has about 73% of white pixel. The RSS 2 statistical result shows that RSS 2 has a range of 34-72% white pixel. Comparison of RSS 1 and RSS 2 result shows that the system works properly that the RSS 2 has less white pixel or has more black pixel which represents more rust and mould.

Table 1. Black and white pixel percentage statistical result compared to actual rust and mould percentage

RSS grade	rust & mould	black pixel	white pixel
RSS 1	0%	$0 - 27\%$	$73 - 100\%$
RSS 2	5%	$28 - 66\%$	$34 - 72\%$
RSS 3	10%	$> 67\%$	$< 34\%$

The system is then tested to recognize a hundred random known sheets. The result shows that from a hundred sheets there are twelve false recognitions. Among twelve false recognitions, there is only one fault on RSS 1 recognition, none of them is on RSS 3 and eleven faults are on RSS 2 . This means that the system is 100% accurate in recognizing the RSS 3. The system is also almost perfect in recognizing RSS 1. The real problem is on RSS 2 recognition which produces eleven faults out of forty one sheets (randomly forty one sheets of RSS 2 is selected to be tested). Most faults happened are the RSS 2 sheet is graded as RSS 1 sheet. One sample of the fault is shown in Figure ??. From the image, it can be seen that the illumination affects the system. There are some area with darker colour in the edge of the image and there are some are with brighter image on the center of the image straight to the camera. This illumination unequal illumination distribution produces difference noise which can not be tackled by statistical threshold. This weakness can be solved by conditioning the environment area and adjusting illumination distribution. Even with the faults, in RSS 2 recognition, the system still obtain a good result with 73% of accuracy.

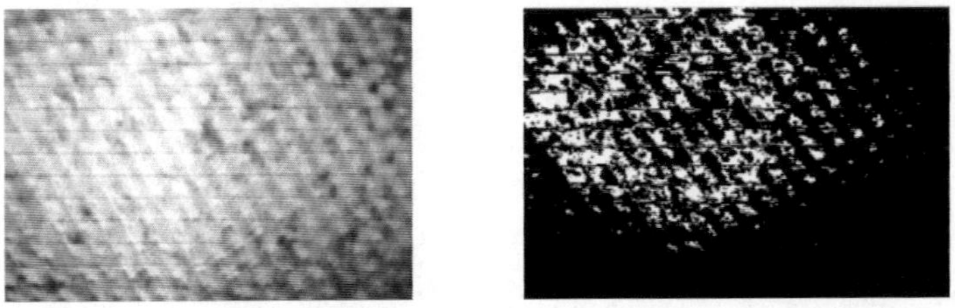

(a) **(b)**

Figure 4. A sample of RSS 2 image with fault recognition in (a) RGB model captured by camera, (b) Binary model after system process

5. Conclusion

The statistical result produces two threshold value to distinguish the RSS grade namely 34% and 73% white pixel on the image. The system can work properly in overall RSS grade recognition with 88% of accuracy. Most faults are found on RSS 2 recognition. This fault is exist due to illumination effect to the image which produce noise or black pixel which can not be tackled by the system statistical threshold. On the next research there must be a designed system which can assure the equal distribution of illumination over the RSS image so that the noise can be decreased significantly.

References

[1] Suseno R S 1989 *Pedoman Teknis Pengolahan Karet Sit yang Diasap (Ribbed Smoked Sheet)* (Bogor, Indonesia : Balai Penelitian Perkebunan Bogor).
[2] Pornpanomchai C and Chantharangsikul N 2010 *Proc. Int. Conf. on Electronics and Information Engineering (Kyoto, Japan)* **1** 128.
[3] Feng G and Qixin C 2004 *Proc. of 5th World Congress on Intelligent Control and Automation* (China) **5** 4802.
[4] Guang-rong L 2010 *Proc. of Int. Con. on Advanced in Energy Engineering* (China) 134
[5] Wang H et al 2014 (China) 5 4802
[6] Cho B H, Bae J W, Jung S H 2008 *Proc. of Int. Con. on Advanced Language Processing and Web Information Technology* (China) 245
[7] Premal C E and Vinsley S S 2014 *Proc. of Int. Con. on Circuit, Power and Computing Technologies* (India) 1229
[8] The Rubber Manufacturer Association, Inc 1979 *The Green Book-International Standard for Quality and Packing for Natural Rubber Grades* (Washington) 5

Mt. Kelud haze removal using color attenuation prior

Fresy N[1,2,a], Eko M Yuniarno[1,b], Supeno M S N[1,c] and M Hariadi[1,d]

[1]Electrical Engineering Department, Faculty of Industrial Technology, Sepuluh Nopember Institute of Technology (ITS), Surabaya, Indonesia
[2]Information Technology Department, Faculty of Science and Technology, Islamic State University (UIN) Maulana Malik Ibrahim, Malang, Indonesia

Email: [a]fresy14@mhs.ee.its.ac.id, [b]ekomulyanto@ee.its.ac.id, [c]mardi@ee.its.ac.id, [d]mochar@ee.its.ac.id

Abstract. Kelud crater observation using closed-circuit television (CCTV) has not been used as the main guide in the world of volcanology. This is caused by observations manually by volcanologist who is not certain and depends on their ability and experience. In practice, there is still obstacles haze in the image taken from CCTV record. This paper present color attenuation prior method to eliminate haze on the digital image. The results obtained showed that the selected method is capable of eliminating sparse haze and moderate haze but not dense haze.

1. Introduction

Concerning about the Mt. Kelud eruption has become central issues in preparedness and mitigation, as a result, much research in recent years has focused on the development of early warning system to observe pre-eruption activity. One of the most important parameters to be observed is the crater lake. Researchers have studied and improved many aspects of the crater lake monitoring at Mt. Kelud. Hidayati et al.[1] responded to the need for detection before the eruption based on color changes to determine the distribution of the crater lake hiposentrum, and Bernard [2] based on water sampling directly of the crater lake, Nandaka and Agung [3] has added a device such as CCTV to observe the smoke activity before eruption. These phenomena known as high activity in the crater lake. It occurs when a perfect mixture of gases (SO_2, HCl, HF and CO_2) from the bottom of the crater lake is strong enough to penetrate the depths of the lake, until it pops out of the lake in the form of a puff of smoke [4]. In this case, the observations of the smoke can be performed using pattern recognition of images taken from the permanent CCTV which is mounted on the side of the volcano's crater lake. Pattern recognition is an objective statistical approach that used for geophysical problems [5,6] search for general patterns in precursory activity before the eruption (e.g., smoke)[7]. However, the resulting CCTV images are visually ineffective due to the haze along the day and it's different thickness. Therefore, this paper proposes a method to remove haze in a single image taken from CCTV, before detecting smoke in the digital image.

During the last twenty years, haze removal techniques were successfully elaborated [8-27]. Initial attempts focused on contrast enhancement of a single image using histogram equalization [8-10]. These methods are time-consuming and are often technically difficult to perform. Some improved algorithms[11-13] suggested a possible instant haze removal of multiple images based on polarization. Unfortunately, these methods do not consider different weather conditions. Therefore, Narasimhan et al. improved haze removal approaches for different weather [14-16]. Tan [17] refined visibility in bad weather from a single image based on Markov Random Field.

In terms of haze removal of color images, [18] proposes independent component analysis (ICA), however, it is difficult to deal with dense-haze images. He et al. [19] suggest the dark channel prior (DCP) approach for non-sky images based one color channel. There remains a need for an efficient method that can handle sky images for color images. Zhu et al. [20-21] propose color attenuation prior for single image haze removal based on a supervised learning method to create a linear model for the scene depth of the hazy image. With the recovered depth information, it can easily remove the haze from a single hazy image.

The purpose of this study is to examine the color attenuation prior to remove haze in a hazy single image which is taken using CCTV. The main body of this paper is divided into three sections. Section 2 consists of a short literature survey. Section 3 begins to apply of the color attenuation prior method at the crater lake single image. Section 4 describes and discusses the result.

2. Haze Removal

Haze removal is a challenging area of research. Among others, Tan [22] observation stated that the haze-free image more contrast than the image containing haze, and be able to maximize the contrast in a local region of the input image. Fattal [23] were able to achieve good results by assuming that the transmission and shadows on the surface locally unrelated. Finally, a very good approximation filed He et al., in which they use dark pixels in the local window dark pixels to produce a rough estimate of the transmission map [24] that obtain results exceeding other algorithms and even managed to dense haze. Several refinement algorithms [25–31], [40–46] are suggested to cope with the shortcoming of the DCP approach. For more efficiency, Gibson et al. [26], Yu et al. [27], He et al. [38], Tarel and Hautiere [40], and Tarel et al. [41] substitute the slow process of soft matting [39] with median filtering, "median of median filter", guided joint bilateral filtering [32–37] and guided image filtering, successively.

2.1. Haze Model

The visual field of computers and computer graphics, mathematical models used to describe the formation of haze is: [14,15,18,21]

$$\boldsymbol{I}_{(x)} = \boldsymbol{J}_{(x)} t_{(x)} + \boldsymbol{A}(1 - t_{(x)}) \tag{1}$$

Where $\boldsymbol{I}_{(x)}$ is the image observed, $\boldsymbol{J}_{(x)}$ is part of the lighting (haze), $\boldsymbol{A}_{(x)}$ is the light of the global atmosphere and $t_{(x)}$ is the portion of light that reaches the camera. $(x) = (x, y)$ is a 2D vector correspond to the coordinates (x, y) of a pixel position in the image. The purpose of detection is to eliminate noise $t_{(x)}$ where $0 < \boldsymbol{I}, \boldsymbol{J}, \boldsymbol{A}, t_{(x)} < \infty$. As shown in Figure 1.

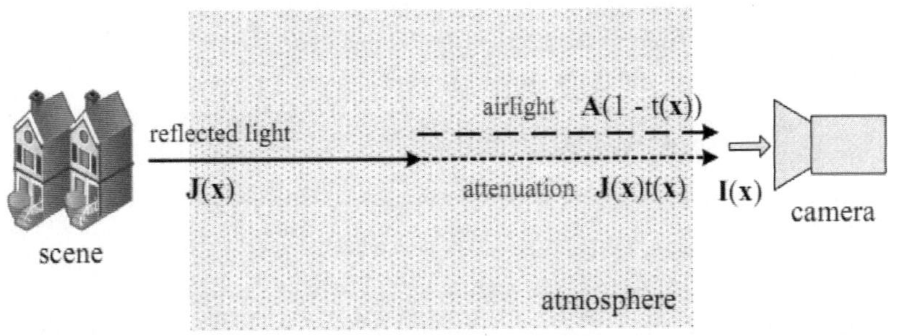

Figure 1. A macro physical picture of the haze imaging model.[14]

In the mathematical model $\boldsymbol{J}_{(x)}\, t_{(x)}$ is called direct attenuation, whereas $\boldsymbol{A}\ (1-t_{(x)})$ describes the airlight [15]. Attenuation directly describes the lighting section which causes damage to the image, while the airlight generated from the scattered light and causing the image of the color shift. At the moment the atmosphere is homogeneous, then the transmission of $t_{(x)}$ can be modeled by: [21]

$$t_{(x)} = e^{-\beta d(x)} \tag{2}$$

Where β is a coefficient indicating the spread of the atmosphere on the lighting value exponentially with the depth of the image *(d)*. Geometrically, in equation (2), may imply that in the RGB-type image, vector $\boldsymbol{A}_{(x)}$, $\boldsymbol{I}_{(x)}$ and $\boldsymbol{J}_{(x)}$ is coplanar and end point parameters are collinear. Transmission $t_{(x)}$ is the ratio of the two line segments:

$$t_{(x)} = \frac{\left\|\boldsymbol{A}_{(x)} - \boldsymbol{I}_{(x)}\right\|}{\left\|\boldsymbol{A}_{(x)} - \boldsymbol{I}_{(x)}\right\|} \tag{3}$$

2.2. *Color Attenuation Prior*

It can be concluded from this context that the effect of the white or gray airlight on the observed values is additive. Thus, caused by the airlight, the brightness are increased while the saturation is decreased. It allows us to take advantage of the difference between the brightness and saturation to estimate the concentration of the haze. Since the haze concentration increases along with the change of the scene depth, then the depth of the scene is positively correlated with the concentration of the haze. This relation can be regarded as color attenuation prior, and written as: [20,21]

$$d(\mathrm{x}) = \theta_0 + \theta_1 v(\mathrm{x}) + \theta_2 s(\mathrm{x}) + \varepsilon(\mathrm{x}) \tag{4}$$

where x is the position within the image, d is the scene depth, v is the brightness component of the hazy image, s is the saturation component, θ_0, θ_1, θ_2 are the unknown linear coefficients, $\varepsilon(\mathrm{x})$ is a random variable representing the random error of the model, and ε can be regarded as a random image.

2.3. *Image Quality Assessment*

Image quality assessment conducted by an objective assessment by using the amount of *Mean Square Error* (MSE) and *Peak Signal to Noise Ratio* (PSNR), both of these quantities are used to compare the pixels at the same position of the two different images. MSE is the average squared error between the original image with the image processing result which can be formulated as follows:

$$MSE = \frac{1}{MN}\sum_{x=0}^{M-1}\sum_{y=0}^{N-1}\left|\left(f(x,y) - g(x,y)\right)^2\right| \tag{5}$$

PSNR is the value ratio between the maximum value color image on filtering results with the quantity of interference (noise), which is expressed in decibels (*dB*), noise is the average root mean squared error $(MSE)^{1/2}$. Mathematically, PSNR can be formulated as follows:

$$PSNR = {}^{10}\log 10\left(\frac{255}{(MSE)^{1/2}}\right) \tag{6}$$

3. Method

Tests performed on a single image of the crater of Mt. Kelud, which was taken in May 2016 belonged to the meteorology agency in East Java, with a size of 800 x 450 pixels. The single image of the crater of Mt. Kelud consist of a sparse hazy image, a moderate hazy image and a dense hazy image. Each performed at different β values, namely $\beta_1 = 1.1$, $\beta_2 = 1.2$, $\beta_3 = 1.4$, $\beta_4 = 1.6$, $\beta_5 = 1.8$ and $\beta_6 = 1.9$.

In figure 2 it can be seen that the color of single images, on the CCTV crater lake, uses the RGB model to determine color. Once the camera has read these values, they are transformed to HSV values.The HSV model depicts colors similarly as the human eye tends to perceive color. RGB defines color in terms of a combination of primary colors, while, HSV delineates color using common comparisons such as color, vibrancy and brightness.

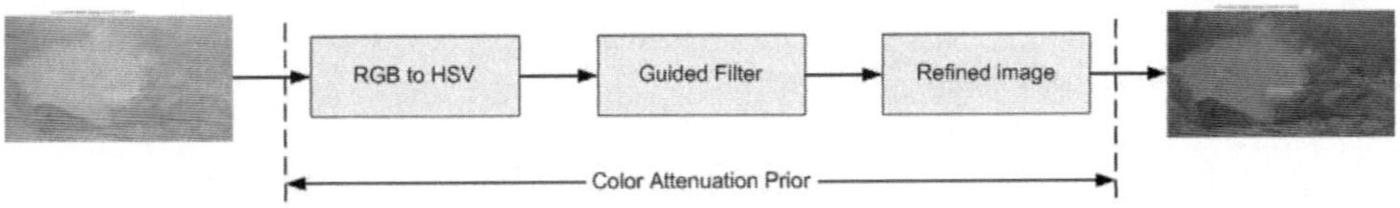

Figure 2. The left image is the hazy image before color attenuation prior method, and the right is the result of haze removal.

Hue represents the color type. Saturation represents the vibrancy of the color. The value represents the brightness of the color. For example, white has an HSV value of 0-255, 0-255, 255. Black has an HSV value of 0-255, 0-255, 0. The HSV values are then used in the code to incorporate additional information from a given guidance image during the filtering process. The filtering output is locally a linear transform of the guidance image. Hereinafter, the filtering output is refined.

4. Results and Discussion

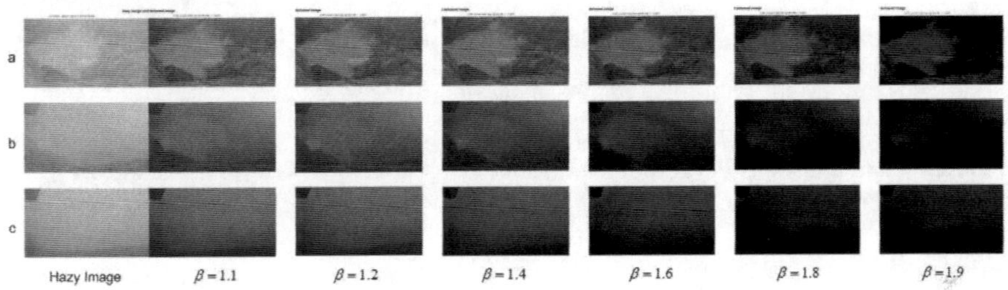

Figure 3. The left image is the hazy image before color attenuation prior method, and the right is the result of haze removal in the case of $\beta_1 = 1.1$, $\beta_2 = 1.2$, $\beta_3 = 1.4$, $\beta_4 = 1.6$, $\beta_5 = 1.8$ and $\beta_6 = 1.9$. (a) A sparse hazy image, (b) a moderate hazy image and (c) a dense hazy image, respectively.

From figure 3 it can be seen that the greater the value of β, the more dark haze removal results obtained. In these trials, for the sparse haze image (a), the beta value of which still look natural, $\beta_1 = 1.1$, $\beta_2 = 1.2$, $\beta_3 = 1.4$, $\beta_4 = 1.6$. For the moderate haze image (b), the value of β for the natural results of haze removal is $\beta_1 = 1.1$, $\beta_2 = 1.2$, $\beta_3 = 1.4$,. Meanwhile, For the dense haze image (c), using the same beta value, the method can not remove the haze. Based on these trials, the proposed method, capable of eliminating the sparse haze, and the moderate haze only but not the dense haze.

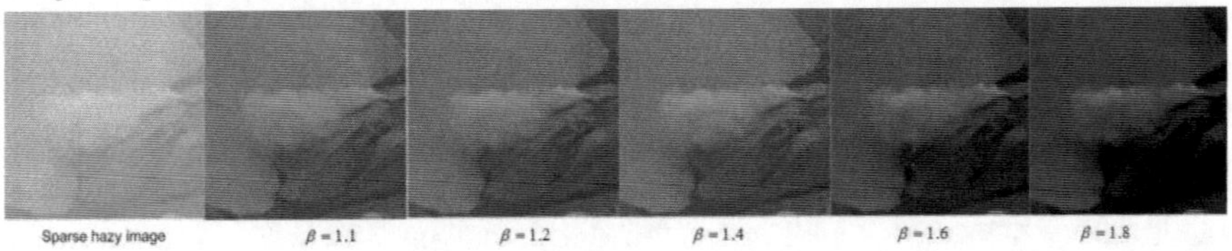

Figure 4 A closer look at the sparse hazy image after haze removal in the case of $\beta_1 = 1.1$, $\beta_2 = 1.2$, $\beta_3 = 1.4$, $\beta_4 = 1.6$ and $\beta_5 = 1.8$. The left image is the hazy image before color attenuation prior method.

Figure 4 gives, a comparison of the result of the haze removal using different β for the sparse hazy image. The natural result is for β = 1.1, β =1.2, β = 1.4. Whereas for β =1.6 and β =1.8, the image less bright than the beta value which is previously mentioned.

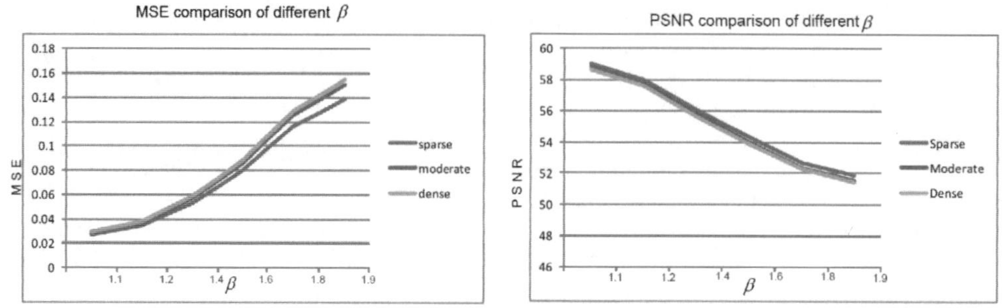

(a) (b)

Figure 5 A comparison of (a) MSE and (b) PSNR in the case of β_1= 1.1,β_2=1.2, β_3=1.4, β_4=1.6, β_5 =1.8 and β_6 =1.9.

From figure 5, it can be seen that the higher β value, the MSE (a) is also increasing. Although the line for moderate haze and dense haze coincident but the results are different. As shown in figure 3. Meanwhile, the higher *β* value, the PSNR (b) is decreasing.

5. Conclusion

In this study, we have applied the color attenuation prior method to remove the haze of the single image taken from Mt. Kelud crater lake. The preliminary results show that this method works properly for sparse haze and moderate haze. The drawback of this method is it can not work properly for dense haze. In the future, we will do further investigation in order to enhance the ability of this method.

References

[1] Hidayati S, Basuki A, Kristianto and Mulyana I December 2009 *Jurnal Geologi Indonesia***4** No. 4 229-38.
[2] Bernard A 25th to 31st July 2013 IAVCEI Commission of Volcanic Lakes 8th *Workshop onVolcanic Lakes*, Aso volcano and Hokkaido Island, JAPAN.
[3] Nandaka and Agung I G M May 24th-28th 2015 *Japan Geoscience Union Meeting*, Makuhari, Chiba, Japan.
[4] Pasternack G B and Varekamp J C 1997 *Bull. Volcanol* **58** 528–38.
[5] Sandri L, Marzocchi W and Zaccarelli L 2004 *Bull. Volcanol* **66** 263–75.
[6] Sandri L, Marzocchi W and Gasperini P 2005 *Geophys. J. Int.***163** 1203–18.
[7] Rosas A T M and Reyna S D C 2008 *J. Volcanol. Geotherm. Res.* **176** 277–90.
[8] Kim T K, Paik J K, and Kang B S Feb 1998 *IEEE Trans. Consum. Electron.* **44** no. 1 pp. 82– 7.
[9] Stark J A May 2000 *IEEE Trans. Image Process.* **9** no. 5 pp. 889–96.
[10] Kim J Y, Kim L S and Hwang S H Apr 2001 *IEEE Trans. Circuits Syst. Video Technol.* **11** no. 4 pp. 475–484.
[11] Schechner Y Y, Narasimhan S G and Nayar S K 2001 *Proc. IEEE Conf. Comput. Vis. PatternRecognit. (CVPR)* pp. I-325–I-332.
[12] Shwartz S, Namer E and Schechner Y Y 2006 *Proc. IEEE Conf. Comput. Vis. Pattern*

Recognit.(CVPR) **2** pp. 1984–91.
[13] Schechner Y Y, Narasimhan S G and Nayar S K 2003 *Appl. Opt.* **42** no. 3 pp. 511–525.
[14] Narasimhan S G and Nayar S K Jun 2000 *Proc. IEEE Conf. Comput. Vis. Pattern Recognit.(CVPR)* pp. 598–605.
[15] Nayar S K and Narasimhan S G Sep 1999 *Proc. IEEE Int. Conf. Comput. Vis. (ICCV)* **2** pp. 820-27.
[16] Narasimhan S G and Nayar S K Jun 2003 *IEEE Trans. Pattern Anal. Mach. Intell.* **25** no. 6 713–24.
[17] Tan R T Jun 2008 *Proc. IEEE Conf. Comput. Vis. Pattern Recognit. (CVPR)* pp. 1–8.
[18] Fattal R Aug 2008 *ACM Trans. Graph.* **27** no. 3 p. 72.
[19] He K, Sun J and Tang X Dec 2011 *IEEE Trans. Pattern Anal. Mach. Intell.* **33** no. 12, pp. 2341–53.
[20] Zhu Q, Mai J and Shao L 2014 *Proc. Brit. Mach. Vis. Conf. (BMVC)* (Nottingham, U.K) pp. 1–10.
[21] Zhu Q, Mai J and Shao L November 2015 *IEEE Trans. Image Process.* **24** No. 11 pp. 3522-33.
[22] Tan R 2008 *Proc. IEEE Conference* on *Computer Vision and Pattern Recognitio CVPR 2008* Jun. pp. 1 –8.
[23] Fattal R August 2008 *ACM Transactions on Graphics* **27** no. 3.
[24] He K, Sun J and Tang X 2009 *Proc. IEEE ComputerSociety Conference on Computer Vision and Pattern Recognition* **0** pp. 1956–63.
[25] Pei S C and Lee T Y Sep/Oct 2012 *Proc. 19th IEEE Conf. Image Process. (ICIP)* pp. 957–960.
[26] Gibson K B, Vo D T and Nguyen T Q Feb 2012 *IEEE Trans. Image Process.* **12** no. 2 pp. 662–73.
[27] Yu J, Xiao C and Li D Oct 2010 *Proc. IEEE 10th Int. Conf. Signal Process. (ICSP)* pp. 1048–52.
[28] Pulung N A, Eko M Y, M Hariadi and Valentijn V May 10-12 2012 *International Conference On Multimedia Computing and System (ICMCS'12)* (Movenpick, Tangier, Morocco).
[29] Zhu Q, Yang S, Heng P A and Li X Dec. 2013 *Proc. IEEE Conf. Robot. Biomimetics (ROBIO)* 1796–00.
[30] Xiao C and Gan J Jun. 2012 *Vis. Comput.* **28** nos. 6–8 pp. 713–721.
[31] Xiang Y, Sahay R R and Kankanhalli M S Jul. 2013 *Proc. IEEE Conf. Multimedia ExpoWorkshops (ICMEW)* pp. 1–4.
[32] Tomasi C and Manduchi R Jan 1998 *Proc. 6th Int. Conf. Comput. Vis. (ICCV)* pp. 839–46.
[33] Paris S and Durand F 2006 *Proc. Eur. Conf. Comput. Vis.* pp. 568–80.
[34] Porikli F Jun 2008 *Proc. IEEE Conf. Comput. Vis. Pattern Recognit. (CVPR)* pp. 1–8.
[35] Yang Q, Tan K H and Ahuja N Jun 2009 *Proc. IEEE Conf. Comput. Vis. Pattern Recognit.(CVPR)* pp. 557–64.
[36] Adams A, Gelfand N, Dolson J, and Levoy M 2009 *Proc. ACM SIGGRAPH* pp. 21:1–21:12.
[37] Adams A, Baek J and Davis M A May 2010 *Comput. Graph. Forum* **29** no. 2 pp. 753–62.
[38] He K, Sun J and Tang X Jun 2013 *IEEE Trans. Pattern Anal. Mach. Intell.* **35** no. 6 pp. 1397–09.
[39] Levin A, Lischinski D and Weiss Y Feb 2008 *IEEE Trans. Pattern Anal. Mach. Intell.* **30** no. 2 pp. 228–42.
[40] Tarel J P and Hautiere N Sep/Oct 2009 *Proc. IEEE 12th Int. Conf. Comput. Vis. (ICCV)* pp. 2201–08.
[41] Tarel J P, Hauti0re N, Caraffa L, Cord A, Halmaoui H and Gruyer D Apr 2012 *IEEE Intell.Transp. Syst. Mag.* **4** no. 2 pp. 6–20.
[42] Ancuti C O, Ancuti C, Hermans C and Bekaert P 2010 *Proc. Asian Conf. Comput. Vis. (ACCV)* 501–14.
[43] Prapti K N, Eko M Y, Supeno M S N 12 September 2015 *Seminar Nasional Pendidikan Teknik Informatika (SENAPATI 2015)* ISSN 2087-658 (Denpasar, Bali).

[44] Nishino K, Kratz L and Lombardi S Jul 2012 *Int. J. Comput. Vis.* **98** no. 3 pp. 263–78.
[45] Meng G F, Wang Y, Duan J, Xiang S and Pan C Dec 2013 *Proc. IEEE Int. Conf. Comput. Vis.(ICCV)* pp. 617–24.
[46] Tang K, Yang J and Wang J Jun 2014 *Proc. IEEE Conf. Comput. Vis. Pattern Recognit.(CVPR)* pp. 2995–02.

Ornament Problem Suppression in Indonesian License Plate Recognition Systems

Samuel Mahatmaputra Tedjojuwono

School of Information Systems, Bina Nusantara University, Jakarta, Indonesia

smahatmaputra@binus.edu

Abstract. Based on the original work of fast performance algorithm in detecting Indonesian license plate, the proposed work will solve the error found in the license plate localization process caused by plate like pattern within the image, which was called the ornament problem. Although not in all cases, this problem could exist when a car has banner, regular pattern, car's front grill, that could miss understood by the system as license plate letters. The proposed work will implement filtering systems instead of machine learning approach. The filtering methods will follows three steps: detection filter based on the number of elements in the vector, based on the letter proportion of a license plate number, and based on the distance between detected letters. This approach will maintain the fast properties of the original algorithm and will increase the accuracy of localizing the license plate within the given image. Keywords—license plate, localization, computer vision.

1. Introduction

In implementing ALPR (automatic license plate recognition) on a real traffic the algorithm implemented needs to be agile and fast. One of the processes of implementing ALPR system is the license plate localization process. The proposed work is based on the previous research that implements fast performance non-learning based algorithm [1]. However the algorithm still suffers from what was proposed in the related work as the ornament problem. The problem exist where contour pattern that has high similarity to a plate alphabet and number. This could negatively impact the license plate localization process. Figure 1 shows the result of wrongly localized license plate as a result of the front grill of the minibus mistakenly translated as the license plate letters. This is as a result of the vehicle's grills contours (shows in figure 2) have high similarity as the basic contours of a series of license plate letters. The proposed work presents an algorithm to suppress this ornament problem in LPR (license plate recognition) system using interconnected image segmentation and relates them to the proportion of the contours' bounding box.

Figure 1. Falsely localized license plate letters caused by ornament problem

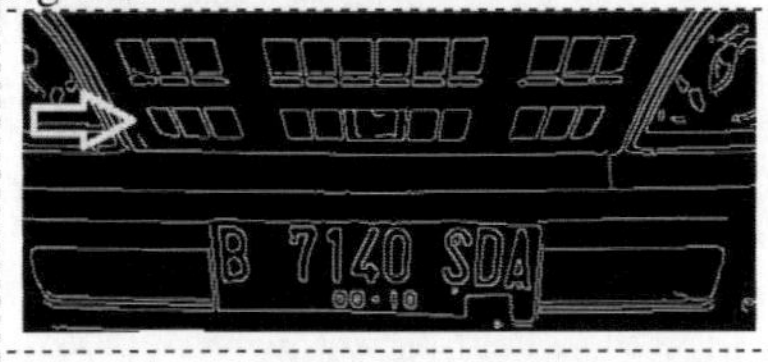

Figure 2. Falsely localized license plate letters described at edge detection image

2. Related work

Recent work was also attempted to solve the problem using the contour detection technique [2] however it was tested against clean sample images that does not contain some of the ornament problems mentioned [1]. Another earlier work [3] used to verify the license plate is by using width and height of license plate candidate, and by implementing the so called horizontal crosscut to count how many objects within the horizontal line. However yet again this approach does not indicate any extreme case that could present in a form of other elements that consist of several objects, which could have high similarity as a license plate - if not higher cost function value. Other used a machine learning approach into what was named as geometric operation [4]. This is where object with certain characteristic were labeled using aspect ratio, total pixel, width, height, and presence of letters. Those aspects then will be fed into a hidden layer using an improved back propagation. The technique and some other previous works [5][6][7] require numbers of samples to improve the localization process.
Rectangle detection in solving plate localization process was implemented using contour detection [8]. However the use of rectangle detection could be misleading to some cases where the background of license plate has similar color value to the vehicle's color, and to the presence of other object in the sample (such as banners) that has rectangular boundary. Another approach as well used the license plate dimension as a method to localize license plate in a sample [9]. This works as well combine it with Euler method to detect the number of close loop of the license plate letters. However it did not tested against sample with more then one high Euler values present in the image which indicating the presence of letters that are not license plate letters in the input sample.

Other recent work shows advance techniques in localizing and filtering various effect of lighting such as shadow, dark condition, exposure, and daylight [10]. However the technique uses local binary method to localize the license plate.

3. The proposed systems

The ornament problem in a LPR systems happen when the localization process fail to detect the location of a license plate within a given image due to distraction from other ornaments of the cars. Figure-3 shows wrong detection marked by the bounding boxes around phone numbers of the car's windows banner. Figure-4 emphasizes the wrong detection by showing the only the bounding boxes of the detected contours image. It displays the bounding box of the location detected by the original algorithm in the 2D Euclidean plane of the given image sample.

Figure 3. Falsely localized license plate letters caused by other letters

Figure 4. Falsely localized license plate letters caused by other letters (bounding box only)

Furthermore, this falsely located license plate was the result of a strong presence of other letters in the sample image that are not part of the license plate letters. The problem usually takes place in a case of vehicle's banners, front grills, logos, and any other regularly placed contours in a real case traffic information systems. The propose systems implement filtering algorithm to suppress the so-called ornament problem [1] by dividing the system into three filtering functions. The three functions could be implemented separately, however through some of the experimental result we could see later on that combining the three will improve the localization process of license plate in Indonesian plate systems.

The proposed algorithm follows the following methods to suppress the ornament problems while still maintaining the fast and accurate properties of the original system:

1. • Detection filter based on the number of elements in the vector
2. • Detection filter based on the letter proportion of a license plate number
3. • Detection filter based on the distance between detected letters.

The first method will filter detected set of contour based on certain threshold. Figure-5 is depicting eleven elements that were considered within the wrong localization process.

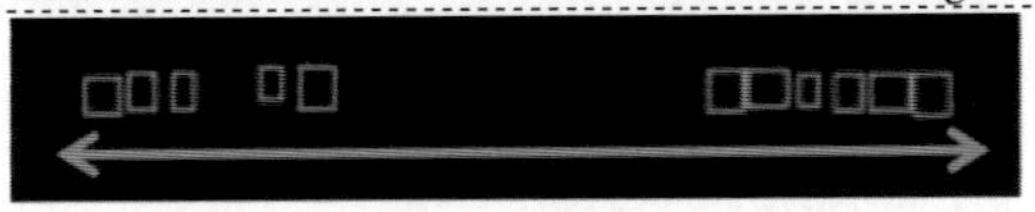

Figure 5. Filter based on counting number of elements within the set

Figure 6. Filter based on ratio of width and height each detected letters

Secondly the algorithm equipped with filtering process based on the proportion of the boxes. A letter within a license plate system will have a certain ratio of width and height of the boxes (figure 6). Lastly, the algorithm proposed to have filtering system based on the distance measured from the center of every box to the center of its neighboring boxes (figure 7).

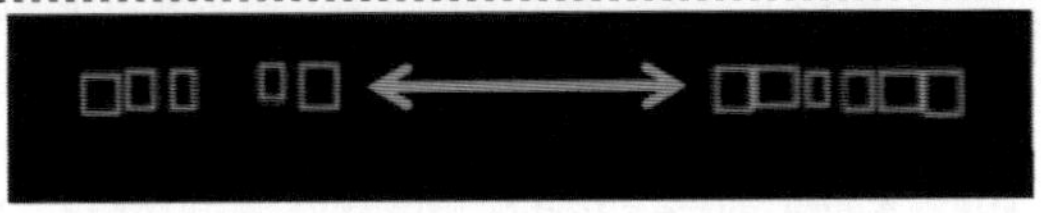

Figure 7. Filter based on measuring the distance of each box with its neighbouring element

Figure 8. Expected result is to move the licence plate falsely located (red arrow)

This filtering mechanism has been added to the original work [1] so it could raise the performance of license plate localization. This method was chosen above other methods that based on machine learning is to keep the fast properties of the original application.

3.1. Expected result

The expected output of the proposed work will be a perfected fast performance algorithm of license plate recognition systems with suppression method for reducing the ornament problem [1] that is exist in a real live traffic monitoring system. Figure-8 shows the expected shift from the back window's banner to the license plate letter below. The algorithm as well expected to translate the localized license plate using optical character recognition technique.

3.2. Filter algorithm by number of letters

The detection filter based on the number of elements within the vector would search for corresponding segmented images that relatively lay at the same horizontal line. Given the vector of segmented images *S*, then each image S_i within S will be compared to its neighbor $S_{j..n}$. If $S_i(x,y)$ position smaller or equal to threshold T of its neighboring segments $S_{j..n}$ than group S_i and $S_{j..n}$ into filtered vector $FS_{i..n}$. For every sub vector FS_i check whether the maximum number of letter is smaller than maximum Indonesian plate number digit MT. Return the largest element in FS. This will be heuristically the strongest candidate as a vector of valid license plate's letters.

3.3. Filter algorithm by letter proportion

To filter using the proportion of width and height of the letter bounding box – this is different from earlier work [3] where the evaluated are was the entire license plate not on each letter, the following pseudo-code was implemented.

```
private boolean validLetterRectangle(Rectangle rectangle){
  Set w= 0.6; // width threshold
  Set h= 0.9;  // height threshold
  R = w/h;
  ratio=(double)rectangle.Width/(double)rectangle.Height;
  if (ratio <= R)
     return true;
  else
     return false
}
```

Figure 9. Filter algorithm using the proportion of a valid letter (pseudo-code)

There are two threshold set for the algorithm based on valid Indonesian license plate letter proportion of width and height, whereas will be called R. This value of R then will be compared to each of the bounding box ratio of width and height $r_{i..n}$ of the segmented images. For every $r_{i..n}$ that is smaller or equal to R, the image within the bounding box will be considered as a valid license plate letter.

3.4. Filter algorithm by distance between letter

This detection filter based on the distance of elements within the vector would search for corresponding segmented images that relatively lay at the same horizontal line and have distance $D_{i..n}$ that is smaller than valid threshold DT. Given the vector of segmented images S, then each image S_i within S will be compared to its neighbor $S_{j..n}$. If $S_i(x,y)$ position smaller or equal to threshold T of its neighboring segments $S_{j..n}$ and D_i is smaller than DT, than group S_i and $S_{j..n}$ into filtered vector $FS_{i..n}$. Furthermore, similar process of examining every sub vector FS_i check whether the maximum number of letter is smaller than maximum Indonesian plate number digit MT.

4. Experimental results

After implementing the filtering algorithms the accuracy of license plate localization process increase by 90%. The following are some of the image resulting from this particular process.

The experiment was tested against 20 various images that were previously have suffered from ornament problem [1]. The time elapsed had also varied depending on the size of the image. The smallest image fed to the systems was at 300x292 pixels with time to process at 65 millisecond. This is including the time to pre-process the image (smoothing), finding contours, filtering the image, and translating them into ASCII character. Proposed algorithm was also tested against smaller license plate image segment size. Figure 10 (a until f) shows the result of experimenting on some plate samples with ornament problems. The accuracy level of this experiment is close to 90% out of the 20 samples.

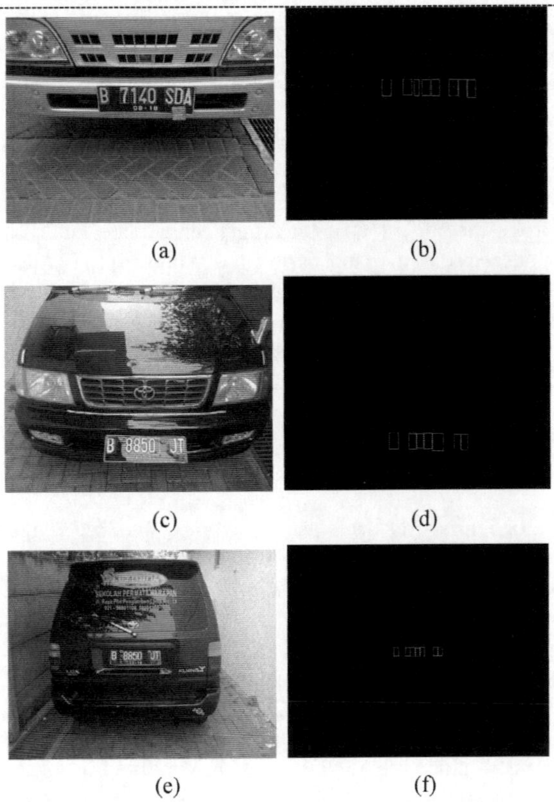

Figure 10. (a until f) Correct experimental result after filtering

As the letter becomes smaller, the contours are not clearly separated each other, and on some cases a letter does not form a perfectly close loop contour. This could lead to missing information that can be seen from figure-11 and figure-12 where both letter "B" at the beginning and the end of the license plate are not detected (located).

Figure 11. Experimental result after filtering on smaller samples

Figure 12. Undetected contours (related to figure 11 g, h) missing information.

The left outer most letter "B" suffer from the merging contour with the underline contour, whereas the most right letter "B" suffer from incomplete close loop contour, and the letter "K" was separated into two close loop contours. These are as result of missing information for small size plate segments' size. To solve this missing information problem, a reasonable size of license plate letter image segments size needed as the input to the ALPR system. Further research is needed to quantify the reasonable size of the input segments in term of pixel values.

5. Conclusion

In conclusion, implementing ALPR on a real live traffic system needs a robust algorithm to be able to handle input irregularity. One form of this arbitrary input is what was proposed to be called as the ornament problem. The algorithm proposed implements several filtering to solve the problem. One of which is to validate the ratio of every image segment bounding box width and height. The result is promising and could improve the accuracy of locating license plate letter within the given input as high as 90%. Moreover, as describe at the later part of the manuscript, an ideal size of the input image to the algorithm is yet to be identified. Future work would need to quantify what could be proposed as the reasonable input size, to minimize the missing information as a result of a "too" small image's segment. Another input irregularity could come in a form of tilted image. Several approaches have been proposed and tested previously. However to the particular algorithm this approach are yet to be proven. Future work needs to identify and solve the tilted image segment problem and combine it with the current approach in order to increase the accuracy of license plate localization method.

References

[1] Tedjojuwono S M 2015 *Fast Performance Indonesian Automated License Plate Recognition Algorithm Using Interconnected Image Segmentation* (Bali: Intelligence in the Era of Big Data) Springer (Berlin: Heidelberg) vol 516 pp 289-300

[2] Kumary R S, Angel B, Laya T 2014 License Plate Detection and Character Recognition Using Contour Analysis *International Journal of Advanced Trends in Computer Science and Engineering* vol **3**, no1

[3] Tran D D, Tran L H, Tran V P, Nguyen V H 2005 *Building an Automatic Vehicle License-Plate Recognition System* (Can Tho, Vietnam: Intl. Conf. in Computer Science) RIVF'05

[4] Nuzulha K I, Emaliana K, Norazira A, Nawawi M R M 2013 License Plate Recognition (LPR): A Review with Experiments for Malaysia Case Study *The International Journal of Soft Computing and Software Engineering [JSCSE]* vol **3** no 3

[5] Chang S L, Chen L S, Chung Y C, Chen S W 2004 Automatic License Plate Recognition *IEEE Transaction on Intelligent Transport System* **5(1)**

[6] Arya F, Suwardi I S 2012 License Plate Recognition System for Indonesian Vehicles *E-Journal Teknologi Industri* **B-87**

[7] Anagnostopoulos C N E, Anagnostopoulos I E, Loumos V, Kayafas E, 2006 A License Plate-Recognition Algorithm for Intelligent Transportation System Applications *IEEE Transaction on Intelligent Transport System* **7(3)**

[8] Gazcón N F, Chesñevar C I, Castro S M 2012 Automatic vehicle identification for Argentinean license plates using intelligent template matching *Pattern Recognition Letters Elsevier* 33 pp 1066–1074

[9] Wazalwar D, Oruklu E, Saniie J 2012 A Design Flow for Robust License Plate Localization and Recognition in Complex Scenes *Journal of Transportation Technologies* vol **2** pp 13-21

[10] Ying W, Yue L, Jingqi Y, Zhenyu Z, Karen M V D, Pengfei S 2011 An Algorithm for License Plate Recognition Applied to Intelligent Transportation System *IEEE Transactions on Intelligent Transportation Systemes* vol **12** no 3

Outer contour extraction of skull from CT scan images

M A Ulinuha[1,2,3], E M Yuniarno[1], S M S Nugroho[1] and M Hariadi[1]

[1] Department of Electrical Engineering, Institut Teknologi Sepuluh Nopember, Surabaya, Indonesia
[2] Universitas Islam Negeri Walisongo, Semarang, Indonesia
[3] Corresponding Author, E-mail: ulinuha@walisongo.ac.id

Abstract. Extraction of the outer contour of the skull is an important step in craniofacial reconstruction. The outer contour is required for surface reconstruction of the skull. In this paper, we propose a method to extract the outer contour of the skull. The extraction process consists of four stages: defining the region of interest, segmentation of the bone, noise removal and extraction of the outer contour based on scanning from the four sides of the image. The proposed method successfully extracts the outermost contour of the skull and avoids redundant data.

1. Introduction

Extraction of the outer contour of the skull is an important step in craniofacial reconstruction [1], [2]. This step is performed before the surface reconstruction to build a 3D model of the head. By finding the contours of the skull, we can perform a contour-based surface reconstruction, which is one classical method of surface rendering that has been developed since 1975 [3] and continues to grow to this day. According to Kumar [4], contour-based surface rendering algorithm is efficient and it reduces the search space, also it decreases the computation cost. This method is simple, easy for calculation and much classical [5].

Contour extraction for surface reconstruction of the skull has its own challenges. On the one hand we have to find the most important contour, but on the other side, the skull structure is very complex. The skull can be divided into two portions: neural and facial. The neural skull is made up of 6 bones: frontal, parietal, temporal, occipital, sphenoid and ethmoid. The facial skull comprises 8 different bones: mandible, maxilla, palatine, zygomatic, nasal, lacrimal, vomer and inferior nasal concha [12]. This complexity make possible to obtain redundant data. Hence it is very important to find the right method so that the contours of the skull can be detected properly without containing a lot of redundant data. In this paper, we propose a method for extracting the outer contours of the skull based on the scanning from the four sides of the image.

2. Related Work

Several studies to extract the contours of the skull been done before. In general, based on the results of the final contour, the detection of the outer contour of the skull can be divided into two, first, methods that detect the outer contour by closing the holes and hollows. The results of these methods are a closed contour. Second, methods that detects the contours of the skull without closing the holes and hollows, so some holes left open. In some parts, the resulting contours are not connected.

In a study [2], active curve is used to extract the contour. The active curve has three forces called extraction force, adhesion force and regularization force. Contours produced from this method are in the form of a single closed contour by filling the holes and hollows that exist in the skull boundary. Curve evolution is used in [6]. The curve used three factors to evolve. These factors named attraction force, adhesive force and regularization force. This research resulted in a single closed contour. The holes or hollows of the skull were closed by the curve. In other study [7], newborn CT images were segmented using coupled level set to detect the inner and the outer contour of the skull. The segmentation results were compared with manual segmentation. The proposed method provided satisfactory result. All of these methods [2], [6], [7], produced a single closed contour, making it less suitable for contour extraction in the part of skull with holes and hollows. Extra contours are added to the holes and hollows, resulting in redundant data.

In a study [8], three steps were performed to extract the outer contour of the skull. The first step was finding contour points using Sobel operator, then a circular scanning was implemented, and then followed by an 8-neighborhood boundary tracing. The results of this study were the outer contour by keeping the holes and hollows. In this method, the user must determine the threshold value which defines the outer contour distance from the center of the image. The threshold usually defined as half of the maximal distance from the skull to the center and the rest must be defined by user manually. This creates difficulties when user works with a lot of data.

The proposed method is designed to avoid redundant data, while minimizing human intervention to the system.

3. Method

The data used in this research is CT scan images of the head of three patients. All data is the result of CT scan of complete head. Data is stored in DICOM format. Each record consists of approximately 300 slices with a size of 512x512 pixels. These slices cut the head in a horizontal position. Thus the CT image we got is a piece of the head when viewed from the top or bottom.

Broadly speaking, the contour extraction method we are proposing consists of 4 stages, namely a) defining the region of interest, b) segmentation of the bone, c) noise removal and d) extraction of the outer contour. Figure 1 shows all the stages of the proposed method.

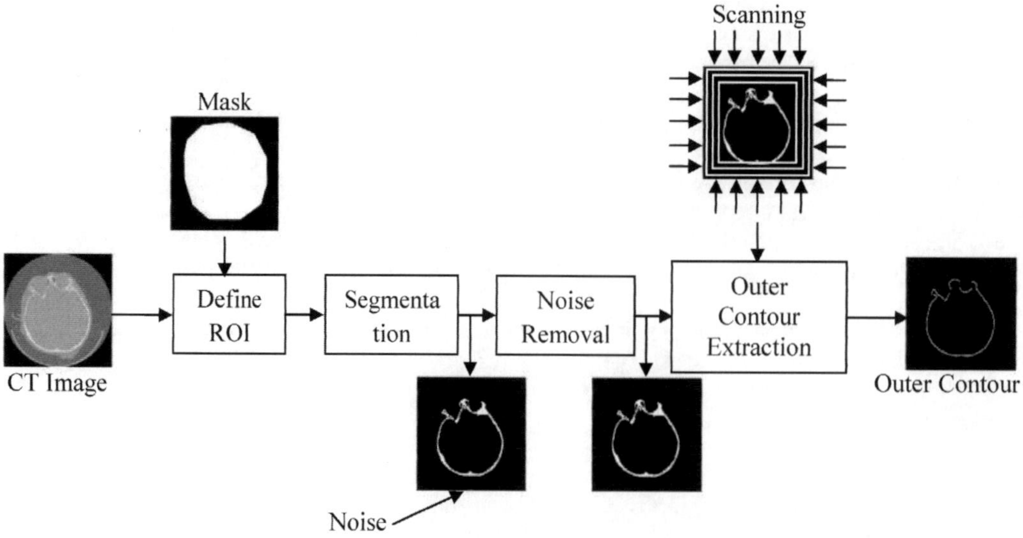

Figure 1. Block Diagam of The Method

The outer contour extraction process is performed per slice. CT scan image is composed of several sections. The part that we want for further analysis is the head. The position of the head is located in the center of the image. To focus on the processing of the head, then we set the center of the image as a region of interest (ROI). Defining ROI is done by a polygon mask which surrounds the head. Polygon mask is in the form of a binary image with a size of 512x512 pixels. Pixels values inside the ROI are 1 and the outside the ROI are 0. This mask works like scissors that cut the part outside of the ROI. By specifying the ROI, then the non-ROI located at the side of the image will be ignored in the subsequent analysis.

If we observe each slice on CT images, the skull bone is brighter than the other regions. In other words, the bone has the highest intensity compared with other regions. Under these conditions, the segmentation of the skull can be done by thresholding. Based on observations of pixel intensity values, the bone has a value of more than 0.8 (pixel values is normalized). Therefore, this value is used as the threshold value. With a fairly high threshold value, only the bones are taken.

Results of segmentation are binary images with pixels values 1 for foreground and 0 for background. It still contains a little noise in the background. This section is obviously not part of the skull because it is located at the lower end of the image. This noise disturbs the contour extraction process, so it must be eliminated. Noise removal can be done easily because the location of the noise can be known with certainty. The noise is removed by replacing the pixels values become equal to the value of the background.

The next stage is to extract the outer contour of the skull. The outer contour is defined as parts of the skull that can be seen from the four sides of the image. Extraction is done by scanning the outer contour starting from the left, right, top and bottom of the image. The scanning process is done in the horizontal direction from the left and right sides, and in the vertical direction from the top and bottom sides. Thus, the number of pixels on the scanning results from left and right sides are the same. Likewise, the number of pixels on the scanning results from the top and bottom. The scanning process is done to find the foreground. Scanning was stopped after the first foreground point is found. The first foreground is the first pixel with value=1 that found in the scanning process. Thus the line which is considered as the outer contour is the entire first foreground pixels resulted in the scanning process.

Results of the outer contour extraction are then compared with the results of contour extraction using Gradient Vector Flow (GVF) snake [9] and Canny Edge Detection [10]. GVF counted 100 iterations by giving a value of μ=0.2. Weights of the snake's internal forces are set at 0.01 to elasticity parameter (α), 3 for rigidity parameter (β) and 1 for the viscosity parameter (γ). Whereas the weight of the external force (κ) was set equal to 2. The initial contour of each slice is the convex hull of the skull bone. Snake deformations performed a total of 250 iterations. Canny Edge Detection is chose because its algorithm performs better than many other operators [11].

4. Result and Discussion

The process of the outer contour extraction of the skull has been successful. The input images are slices of head CT scan with a size of 512x512 pixels. Determination of the region of interest have been successfully separated the image that are not needed in the analysis. Only the head is used for further analysis.

The segmentation separates the bones from the head. The bones are composed of several parts. At slice 1 to slice 89 there are portions of the spine. The bones at slice 90 to 180 shaped like a circle with some breaks. Meanwhile, the shape of the bones at slice 181 to 306 are almost a circle without breaks.

Segmentation is performed by thresholding. It produces noise at the bottom of the image. The noise is very disturbing and must be eliminated. Fortunately the noise position can be known with certainty. From observation, the noise lies at the bottom of the image on the line 510 up to 512. Thus, the pixel values of the lines are converted to 0. Finally, we obtain the skull bone without noise.

After segmentation, outer contour scanning is performed. Visually, the scanning results for slice 1 to slice 180 seem to disconnect some parts of the skull. In the segmentation result, these parts are connected properly. However, because what we want is the outermost contour, the disconnected parts are the best result because the parts will not be visible when viewed from the outside. These parts can be referred to as redundant data. By eliminating redundant data, contour data become smaller and easy to compute.

For slice 181 and the next, contour extraction results are in the form of a single closed contour without disconnected parts. This occurs because the bone structure is approaching a circle, so that no part is lost when viewed from four sides of the image. The extraction process succeeded in eliminating the contours inside the skull. Results of the outer contour extraction process are shown in figure 2.

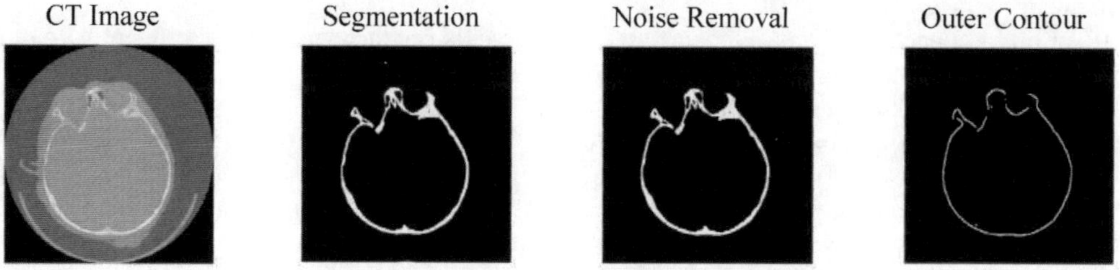

Figure 2. Result of Contour Extraction Process

The outer contour extraction results compared to two other methods, GVF Snake and Canny Edge Detection. From visual observations can be seen that the contours produced by GVF Snake are in the form of a single closed contour by adding contours on some parts that should be disconnected. Results of Canny Edge Detection include the contour inside the skull. While the proposed method managed to get the contour without adding parts that should be disconnected and also eliminates the contour inside the skull. Figure 3 shows the visual differences between the three methods.

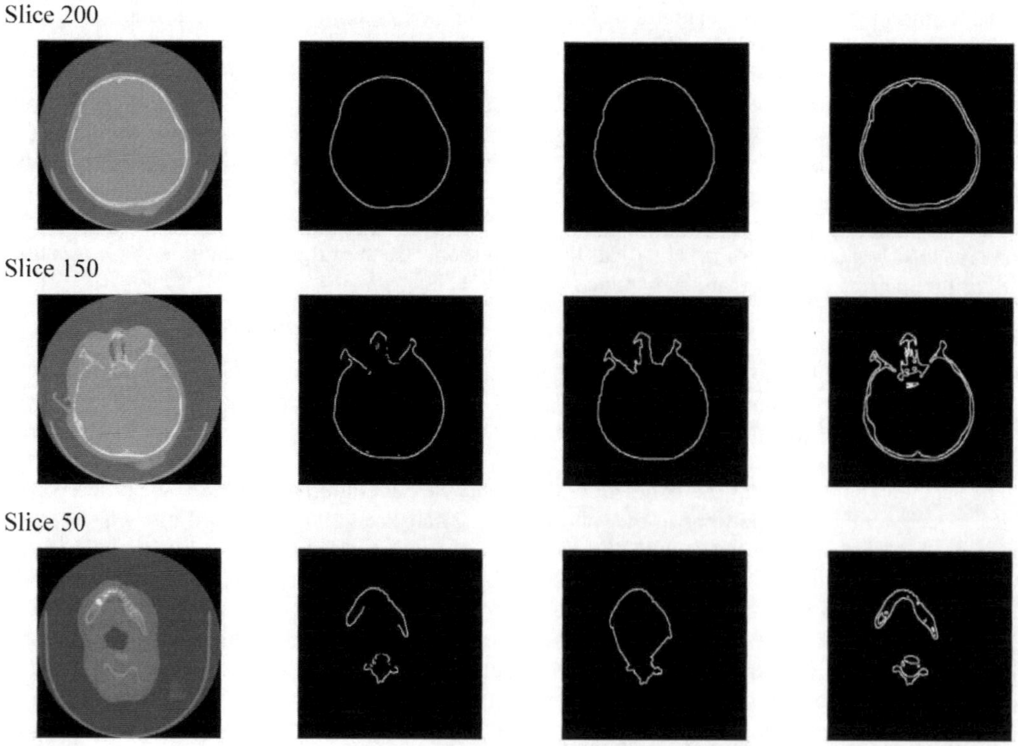

Figure 3. Visual Differences. From left to right: Our method, GVF Snake, Canny Edge Detection

Judging from the thickness of the contours obtained, the proposed method successfully extracts the contour with a thickness of one pixel in all parts. GVF Snake and Canny Edge Detection also detects contour with a thickness of one pixel, but in some parts of the contour, it has a thickness of two pixels. This happens on sloping inclined contour line (instead of vertical or horizontal). In this case, the measurement of the thickness contours conducted in vertical or horizontal direction.

In terms of the number of pixels, the proposed method generates contour with a small number of pixels. It is smaller than pixels extracted by Canny Edge Detection. Thus it can be said that the proposed method only took required pixels. The method managed to eliminate many redundant data that is not needed. But in many slices GVF Snake extract smaller number of pixels. This happens because GVF Snake disconnects some parts of the contour extracted. In fact, this contour should be connected. This case does not happen on the Canny Edge Detection and our method; all contours obtained are connected properly. Figure 4 shows the pixel observation of the three methods. Comparison of the results of contour extraction by the three methods is shown in table 1.

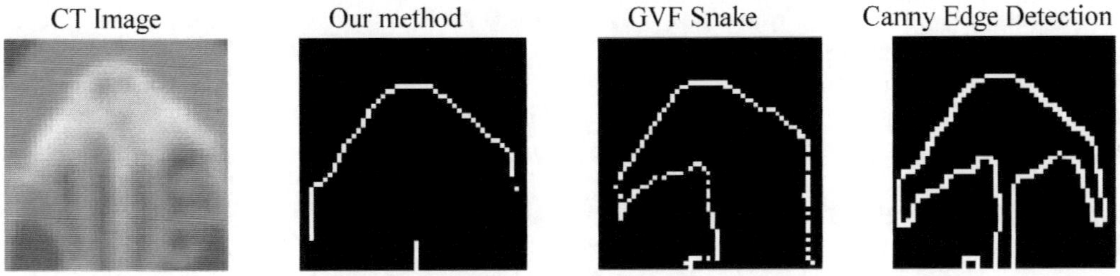

Figure 4. Result of Pixels Observation

Table 1. Comparison of The Methods

Object of Observation	Our method	GVF Snake	Canny Edge Detection
Single contour	Yes	Yes	No
Added contour	No	Yes	No
Redundant contour	No	Yes	Yes
Two pixel contour thickness	No	Yes	Yes
Cut off	No	Yes	No
Number of pixels extracted :			
Slice 250	874	853	2353
Slice 200	978	942	2696
Slice 150	997	1239	3355
Slice 50	567	659	1474

5. Conclusion

We have presented a method for extracting the outer contour of the skull of a CT scan of the head. Contour extraction process begins by selecting a region of interests, followed by segmenting the bones using a thresholding technique. This was followed by noise removal and ends with the extraction of the outer contour. The proposed method has several advantages compared with GVF Snake and Canny Edge Detection. The proposed method avoids adding extra contour, eliminates redundant contours inside the skull and generates contour with a thickness of one pixel. In this case, it can be said that the proposed method is better than GVF Snake and Canny Edge Detection. In the next research we will reconstruct the skull surface based on the contours extracted.

References

[1] Deng Q, Zhou M, Wu Z, Shui W and Ji Y 2016 A regional method for craniofacial reconstruction based on coordinate adjustments and a new fusion strategy *Forensic Sci. Int.* **259** 19-31

[2] Tilotta F, Richard F, Glaunès J, Berar M, Gey S, Verdeille S, Rozenholc Y and Gaudy J F 2009 Construction and analysis of a head CT scan databases for craniofacial reconstruction *Forensic Sci. Int.* **191** no. 1-3

[3] Keppel E Approximating Complex Surfaces by Triangulation of Contour Lines 1975 *IBM J. Res. Dev.* **19** no. 1 2-11

[4] Kumar T S and Vijai A 2012 A 3D reconstruction of a face from 2D CT scan images *Procedia Eng.* **30** no. 2011 970-7

[5] Guan Y G Y, Li Y L Y, Wang Y W Y and Zhang H Z H 2009 Study of Surface Reconstruction Based on Contours of CT Image Slices *Int. Work. Intell. Syst. Appl.* 1-4

[6] Li K, Geng G and Peng S 2014 Single-Layer Closed Contour Extraction from Craniofacial CT Data Using Curve Evolution *Adv. in Intell. Syst. and Computing* **255** 525-32

[7] Ghadimi S, Moghaddam A H, Grebe R and Wallois F 2016 Skull Segmentation and Reconstruction From Newborn CT Images Using Coupled Level Sets *IEEE J. Biomed. Heal. Informatics* **20** no. 2 563-73

[8] Deng Q, Zhou M, Shui W, Wu Z, Ji Y and Bai R 2011 A novel skull registration based on global and local deformations for craniofacial reconstruction *Forensic Sci. Int.* **208** no. 1-3 95-102

[9] Xu C and Prince J L 1998 Snakes, shapes, and gradient vector flow *IEEE Trans. Image Process.* **7** no. 3 359-69

[10] Canny J 1986 A computational approach to edge detection *IEEE Trans. Pattern Anal. Mach. Intell.* **8** no. 6 679-98

[11] Maini R and Aggarwal H 2009 Study and comparison of various image edge detection techniques *Int. J. Image Process.* **3** no. 1 1-11

[12] Gallucci M, Capoccia S and Catalucci A 2007 *Radiographic Atlas of Skull and Brain Anatomy* (Berlin, Heidelberg: Springer Berlin Heidelberg)

IOP Conf. Series: Materials Science and Engineering **185** (2017) 012029 doi:10.1088/1757-899X/185/1/012029

Color Swapping to Enhance Breast Cancer Digital Images Qualities Using Stain Normalization

Izzati Muhimmah[1], Dhina Puspasari Wijaya[2], Indrayanti[3]

Department of Informatics Engineering, Universitas Islam Indonesia[1]

Magister of Informatics Engineering, Universitas Islam Indonesia[2]

Faculty of Medicine and Health Science, Universitas Muhammadiyah Yogyakarta[3]

Jl. Kaliurang Km 14 Yogyakarta 55510

Phone (0274) 895287 ext. 122, fax (0274) 895007 ext. 148

izzati@uii.ac.id[1]* , dhina.puspa@gmail.com[2]

Abstract. Histopathology is the disease diagnosis by means of the visual examination of tissues under the microscope. The virtually transparent tissue sections were prepared using a number of colored histochemical stains bound selectively to the cellular components. A variation of colors comes to be a problem in histopathology based upon the microscope lighting for the range of factors. This research aimed to investigate an image enhancement by applying a nonlinear mapping approach to stain normalization and histogram equalization for contrast enhancement. Validation was carried out in 59 datasets with 96.6% accordance and expert justification.

1. Introduction

Histopathology is a disease diagnosis by the visual examination of tissue under the microscope. To examine the tissue sections (which are virtually transparent), they are first prepared using colored histochemical stains that bind selectively to cellular components[4]. A variation of colors can be a problem in histopathology commonly generated from microscope lighting, different scanners, staining chemical variables/reactivity, and transmission of light as a function of the thickness of the part.

Most of the applications that have been developed for color enhancement in microscopic images at this time are concentrated on the features of shapes and ignore the color between slides. When the color information is utilized, the easiest method is by using the color values obtained from scanning. The most popular staining methods for medical diagnosis are stain hematoxylin (H) and stain eosin (E)[6]. The improvement of image quality is a fundamental process of image processing applications; for images taken under bad conditions, the images often become sharper but with the reduction of the staining in the images[6]. Detail images provide the important information for analysis and thus image enhancement is desired[1].

Histogram equalization methods are often employed to correct low contrast images[3]. Rong & Dongnan (2015) developed a color correction from histogram equalization. They used the greyscale images and their work has never been used in microscopic images with more than one layer. Lin et al. (2015) proposed an approach to stretch the layer colors with histogram averaging and remapping. Their method did not target any enhancement; thus we cannot compare the results for image enhancement standards. An efficient algorithm for image enhancement is needed to improve the image quality in image visualization and image analysis using the computer.

This research investigated a number of methods to enhance image color quality by applying a non-linear mapping approach to stain normalization[4]. We also improved the contrast in the color image of each stain with the histogram equalization of each RGB layer[7].

This research is expected to produce a better model. Testing results have been measured by the distance averages for the RGB layers between the target and purposed images. We aimed to improve breast cancer histology color image by swapping the colors to for the more detail access such as the nucleus and membrane.

2. Materials & Methods

2.1. Color Deconvolution

In the original image, there is a wide, heterogeneous and complex range of colors. Therefore, color separators are used to group the colors of the same type. Ruifrok (2001) mentioned that color deconvolution transforms the red, green, and blue (Ψ) layers into a new space layers of colors red, green, blue ($\widehat{\Psi}$) using a tissue staining color selection. The original (Ψ) and new ($\widehat{\Psi}$) image has three stain color descriptors used to represent an optical density. Ruifrok's (2001) stain has some specific characteristics in the three-layer RGB; each of RGB value is represented in the matrix vector with the optical density of 3x1. The implementation of optical density is symbolized with P. P values include the types of solutions: Hematoxylin (H), Eosin (E) and Diaminobenzine (DAB). Here is a stain for one type of solution:

$$[P_1 P_2 P_3] \tag{1}$$

Normalization P is achieved by dividing each vector of optical density (OD)[8]

$$\hat{P}_{i,j} = p_{i,j} / \sqrt{p_{i,j}^2 + p_{i,j}^2 + p_{i,j}^2} \tag{2}$$

This results in a normalized OD matrix, M:

$$\begin{matrix} \hat{p}_{11} & \hat{p}_{12} & \hat{p}_{13} \\ \hat{p}_{21} & \hat{p}_{22} & \hat{p}_{23} \\ \hat{p}_{31} & \hat{p}_{32} & \hat{p}_{33} \end{matrix} \tag{3}$$

If C is the 3x1 vector for the three stains at a particular pixel, then the vector of the OD levels detected at the pixel is y=CM.

$$C = M^{-1}[y] \tag{4}$$

This implies that multiplying the OD image with the inverse of the OD matrix, defined as color deconvolution matrix D, results in an orthogonal representation of the stains forming the image:

$$C = D[y] \tag{5}$$

The stain value was used to determine the optical density of each layer. In each solution of Hematoxylin, Eosin, and Diaminobenzine, the stain values are unique in each of RGB layer[8].

2.2. Normalization Stain Matrix Algorithm

The proposed algorithm (see Figure 1) consisted of four modules: stain matrix estimation, CD, nonlinear mapping of channel statistics, and reconstruction. There were two input images used, which were served as the source and target images. These images have some different qualities in terms of their stain solution. The nonlinear correction (mapping) was applied to each channel. Then, data were normalized and separated from the target image (based on statistics calculated from the target image).

Finally, this research reconstructed the normalized source image using the normalized stain channel of the target. The following sections present the details of each of these modules.

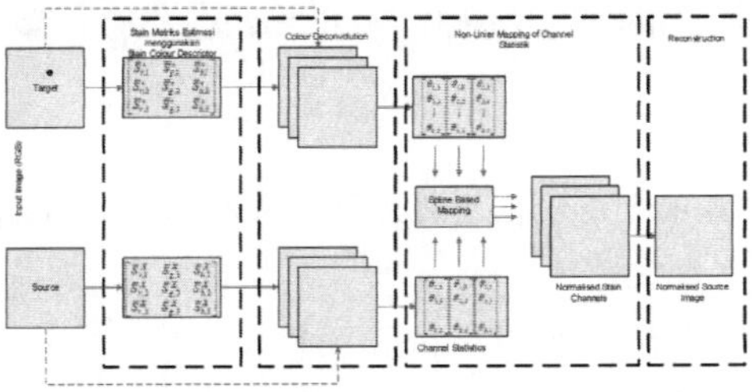

Figure. 1 Normalization Stain Algorithm

2.2.1. Stain Matrix Generation

This study used the CD framework to convert both source and target images from RGB color space to a new one defined by the constituent stains. This required the estimation of image-specific stain matrix (S) each for source and target images. The estimation of image specific stain matrices used a global (preimage) SCD (Stain Color Descriptor) and local pixel-level color information in a supervised color classification framework. Figure 2 shows the diagram of the stain color descriptor process.

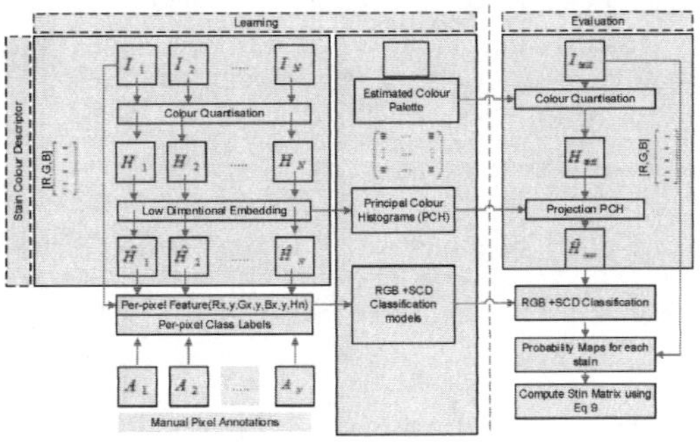

Figure 2 Stain Color Descriptor

Figure 2 presents the overview of the stain matrix estimation method consisting of two phases: learning and evaluation. Learning, performed offline, essentially involved two steps: 1) deriving principal color (PCH) from the training set of quantized image histogram to obtain SCDs and 2) the learning classification models by utilizing RGB (pixel information) and SCD (whole-image color information) in a supervised classification framework to generate the stain-specific probability maps (one for each stain and background). These probability maps were used to estimate the color of each stain for a particular image.

The following rules were used to categorize the images into three classes: background, stained, and other (see Table 1).

Table 1. Rules for categorizing an image into three classes

If	$P(bgd\|\mathcal{F}) > \mathcal{T}_{bgd}\zeta(c)$	=	BACKGROUND
Else if	$P(s_n\|\mathcal{F}) > \mathcal{T}_{fgd}\zeta(c)$	=	STAINED
Else	$\zeta(c)$	=	OTHER

2.2.2. Nonlinear Mapping of Channel Statistics

IOP Conf. Series: Materials Science and Engineering **185** (2017) 012029 doi:10.1088/1757-899X/185/1/012029

Each channel of deconvolved target and source image, the spline-based nonlinear mapping was used to calculate the statistics set and their corresponding probability map of both images. These values were then utilized to map the statistics of each source channel to replace the corresponding target image channel.

2.2.3. Reconstruction

Once each of the stain channels of source image was normalized independently, they were then recombined on each pixel basis as follows:

$$\begin{bmatrix} X_r^{norm}(c) \\ X_g^{norm}(c) \\ X_b^{norm}(c) \end{bmatrix} = \begin{bmatrix} 255\text{X}\prod_{n=1}^{3} e^{-\hat{X}_n^{norm}(c) s_{r,n}^{\mathfrak{I}}} \\ 255\text{X}\prod_{n=1}^{3} e^{-\hat{X}_n^{nor}\ (c) s_{g,n}^{\mathfrak{I}}} \\ 255\text{X}\prod_{n=1}^{3} e^{-\hat{X}_n^{norm}(c) s_{b,n}^{\mathfrak{I}}} \end{bmatrix} \tag{6}$$

where c ∈ C refer to a pixel on a 2-D grid ($C,\hat{X}_n^{norm}$) refers to the normalized stain channel n and $s_{\alpha,n}^{\mathfrak{I}}$ (where α∈{r,g,b}) is the stain vector associated with n channel of the normalized $s^{\mathfrak{I}}$ stain matrix[4].

2.3. Histogram Equalization

Each color pixel was represented by a vector with as many components as the color components in a proper color space (i.e., three components: Red, Green, and Blue in the RGB space). It was enhanced by smoothing the multi-level techniques derived from the statistical language techniques.

Greyscale histogram equalization attempted to uniformly distribute the pixel gray levels of an image to all available gray levels L (e.g. L = 256, when 8 bits were used to represent each gray level). Let consider the image pixel gray level to be an rv x. this histogram of a grayscale image is the probability density function (PDF) of x defined as

$$f_x(x_k) = P\{x = x_k\} = \frac{N(x_k)}{\sum_{m=0}^{L-1} N(x_m)} \quad \forall\, k = 0,1,\dots,L-1 \tag{7}$$

where $N(x_k)$ is the number of pixels with gray level value x_k. The Cumulative Distribution Function, CDF, $f_x(x_k)$ of the rv x is given by

$$y_k = F_k(x_k) = P\{x \le x_k\} = \sum_{m=0}^{k} f(x_m)\ \forall k = 0,1,\dots,L-1 \tag{8}$$

This is the simplest approach to color the histogram equalization. Since many color images have three color bases, the color on each pixel is represented by 3-dimentional vector and greyscale histogram equalization is performed in each of the three color components separately. For color image, the color of each pixel was assumed to be a random vector= (XR, XG, XB), where XR, XG, XB were rvs modeling the Red, Green and Blue components, respectively. Thus, by applying equation (8) the equalized histogram of each color component was estimated[2].

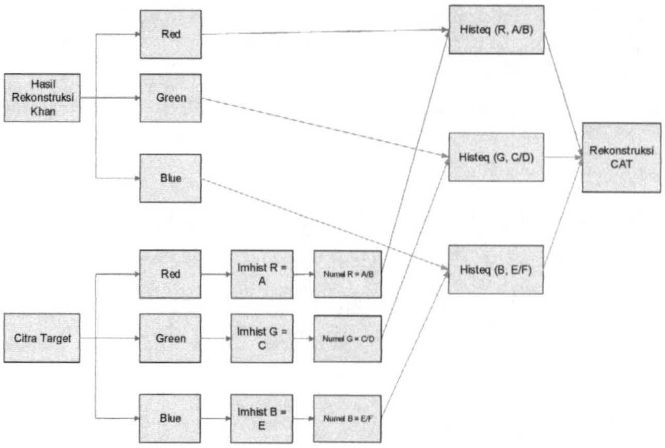

Figure. 3. Chart Research Process

3. Research Methodology

The image data examination of HE breast cancer cells was taken from anatomic pathology laboratory of medicine, Universitas Muhammadyah Yogyakarta. The process of getting the image started from sampling tissue in breast cancer patients seen under the lens Olympus CX31 microscope with a magnification of 40X. Cells were taken in the paint using Hematoxylin (H) and generating substrate Eosin (E). Reaction bonding that occurred between cells and dye caused the cell to be blue and magenta. Image had 1920x1080 dimensions.

Figure 3 presents the research scheme in which an image of the target would map the source images with different color. The reconstruction results in Figure 3 used the method of histogram equalization on each layer RGB.

4. Results

Target image and source image were selected from the image data of HE breast cancer. The target image was as a reference image and the source image was the image that would be color corrected. Figure 3 presents the results of the repair process. Each source image and target image did the method of color deconvolution, and then normalization stain mapping to stain H, stain E, and background. The results reconstructed the normalization stain (Khan normalization method results[4]). The proposed method was conducted by applying the histogram equalization of the image reconstruction results normalized stain.

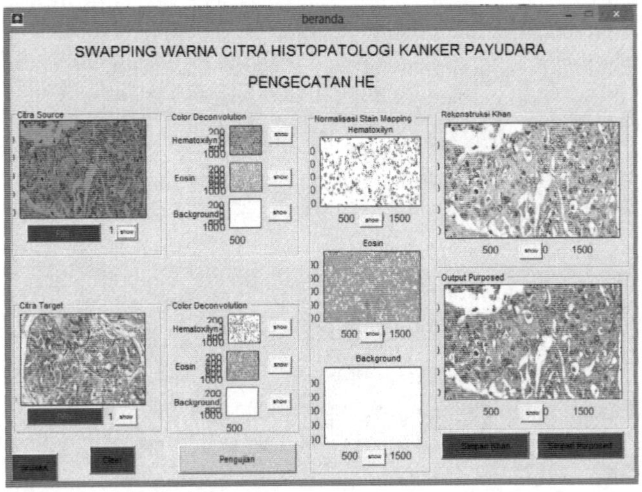

Figure. 4. Figure result color swapping

Tests performed image calculated distance method with pdist2 function in Matlab version 2015b. The result on Khan's method calculated the distance methods resembling to the image target as well as the methods proposed. The closer the image, the more it resembles. The result of the calculation of distance is displayed in Figure. 4.

The evaluation of the quality of the image enhancement was also justified by the pathologist. We had provided the pathologist with 59 sets of source image, target image, Khan's, and the proposed results. We asked her to visually compare which the enhancement results provided the detail information on cell and membranes better. We took the notes on her scoring.

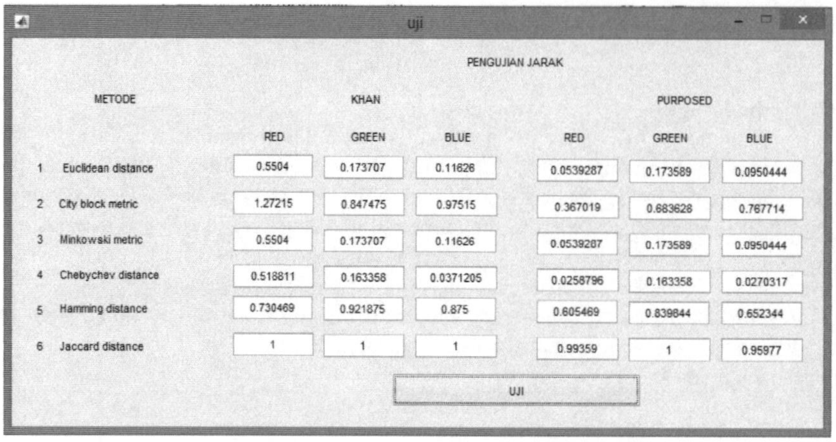

Figure. 5. Results of Distance Testing

Figure 5 shows the results of the average distance of methods Khan[4] for each layer R= 0,770371667 G= 0,546687 B= 0,519965083. On the other hand, the average distances of the proposed method each layer were R= 0,349969167 G= 0,505668 B= 0.43282475. From the calculation of the distance it indicated a distance between the image of the proposed method and the image of the target closer.

Testing models used 59 datasets. The results of proposed method declared 57 dataset in accordance with the criteria of experts. Whilst, there were 2 dataset declared better than Khan's by the experts. The percentage of testing results declared that the proposed method reached an agreement with the expert criteria by 96.6%.

5. Conclusion

This study is the first process (preprocessing) for detecting the presence of breast cancer in the image HE. From the results of the implementation method, it is found that the proposed method has more similarities than previous methods proven by using the method of calculation of the distance.

Acknowledgement

A research grant from PUPT from Ministry RISTEK-DIKTI of Indonesia for the period of 2015-2016, contract no. 001/HB-LIT/III/2015, is gratefully acknowledged.

6. References

[1] Bai X 2015 Morphological Feature Extraction for Detail Maintained Image Enhancement by Using Two Types of Alternating Filters and Threshold Constrained Strategy *Elsevier Optic* **126** pp 5038-43.

[2] Bassiou N and Kotropoulus C 2007 Color Image Histogram Equalization by Absolute Discounting Back-Off *Compt. Vis. Image Underst* **107(1)** pp 108-22.

[3] Jiang G, Lin S F, Wong C Y, Rahman M A, Ren T R, Kwok N, Shi H, Yu Y H and Wu T 2015 Color Image Enhancement with Brightness Preservation Using Histogram Specification Approach *Elsevier Optic* **126** pp 5656-64.

[4] Khan A M, Rajjpoot N, Treanor D and Magee D 2014 A Nonlinear Mapping Approach to Stain Normalization in Digital Histopathology Images Using Image-Specific Color Deconvolution *Biomed. Eng.* Vol 61 **6** pp 1729-38.

[5] Lin S F, Wong C Y, Jiang G., Rahman M A, Ren T R, Kwok N, Shi H, Yu Y H and Wu T 2016 Intensity and Edge Adaptive Unsharp Masking Filter for Color Image Enhancement *Elsevier Optic* **127** pp 407-14.

[6] Macenko M, Niethammer M, Marron J S, Borland D, Woosley J T, Guan X, Schmidt C, and Thomas N E 2009 A Method For Normalizing Histology Slides for Quantitative Analysis *Proc. of the Sixth IEEE Int. Conf. on Symposium on Biomedical Imaging: From Nano to*

Macro pp 1107-10.

[7] Rong Z, Li Z and Dong-nan, L 2015 Study of Color Heritage Image Enhancement Algorithms Based on Histogram Equalization *Elsevier Optic* **126** pp 5665-67.

[8] Ruifrok A C and Johnston D A 2001 Quantification of Histochemical Staining by Color Deconvolution *The international academy of cytology analytical and quantitative cytology and histology* pp 291-9.